T0321782

Sliding Mode in Intellectual Control and Communication:

Emerging Research and Opportunities

Vardan Mkrttchian
HHH University, Australia

Ekaterina Aleshina
Penza State University, Russia

www.igi-global.com

Published in the United States of America by
IGI Global
Information Science Reference (an imprint of IGI Global)
701 E. Chocolate Avenue
Hershey PA, USA 17033
Tel: 717-533-8845
Fax: 717-533-8661
E-mail: cust@igi-global.com
Web site: http://www.igi-global.com

Library of Congress Cataloging-in-Publication Data

Names: Mkrttchian, Vardan, 1950- author. | Aleshina, Ekaterina, 1978- author.
Title: Sliding mode in intellectual control and communication : emerging research and opportunities / by Vardan Mkrttchian and Ekaterina Aleshina.
Description: Hershey, PA : Information Science Reference, [2017] | Includes bibliographical references.
Identifiers: LCCN 2016057762| ISBN 9781522522928 (hardcover) | ISBN 9781522522935 (ebook)
Subjects: LCSH: Knowledge management. | Communication. | Information theory. | Sliding mode control. | Intelligent control systems.
Classification: LCC HD30.2 .M585 2017 | DDC 658.4/013--dc23 LC record available at https://lccn.loc.gov/2016057762

This book is published in the IGI Global book series Advances in Wireless Technologies and Telecommunication (AWTT) (ISSN: 2327-3305; eISSN: 2327-3313)

British Cataloguing in Publication Data
A Cataloguing in Publication record for this book is available from the British Library.

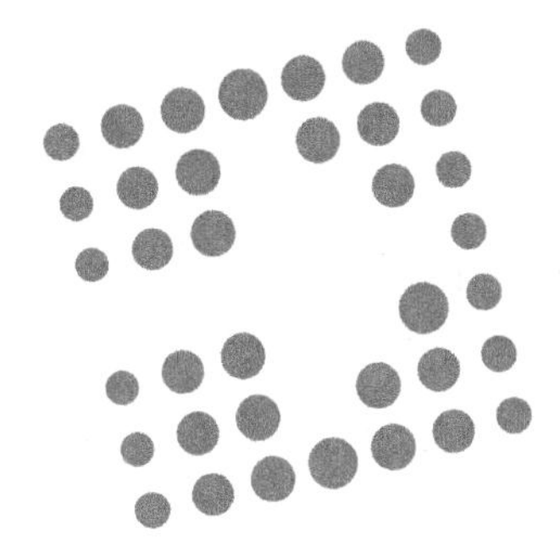

Advances in Wireless Technologies and Telecommunication (AWTT) Book Series

ISSN:2327-3305
EISSN:2327-3313

Editor-in-Chief: Xiaoge Xu, The University of Nottingham Ningbo China, China

MISSION

The wireless computing industry is constantly evolving, redesigning the ways in which individuals share information. Wireless technology and telecommunication remain one of the most important technologies in business organizations. The utilization of these technologies has enhanced business efficiency by enabling dynamic resources in all aspects of society.

The **Advances in Wireless Technologies and Telecommunication Book Series** aims to provide researchers and academic communities with quality research on the concepts and developments in the wireless technology fields. Developers, engineers, students, research strategists, and IT managers will find this series useful to gain insight into next generation wireless technologies and telecommunication.

COVERAGE

- Telecommunications
- Grid Communications
- Network Management
- Radio Communication
- Cellular Networks
- Wireless Broadband
- Wireless Sensor Networks
- Mobile Communications
- Virtual Network Operations
- Digital Communication

IGI Global is currently accepting manuscripts for publication within this series. To submit a proposal for a volume in this series, please contact our Acquisition Editors at Acquisitions@igi-global.com or visit: http://www.igi-global.com/publish/.

The Advances in Wireless Technologies and Telecommunication (AWTT) Book Series (ISSN 2327-3305) is published by IGI Global, 701 E. Chocolate Avenue, Hershey, PA 17033-1240, USA, www.igi-global.com. This series is composed of titles available for purchase individually; each title is edited to be contextually exclusive from any other title within the series. For pricing and ordering information please visit http://www.igi-global.com/book-series/advances-wireless-technologies-telecommunication-awtt/73684. Postmaster: Send all address changes to above address.

Titles in this Series

For a list of additional titles in this series, please visit:
http://www.igi-global.com/book-series/advances-wireless-technologies-telecommunication-awtt/73684

Resource Allocation in Next-Generation Broadband Wireless Access Networks
Chetna Singhal (Indian Institute of Technology Kharagpur, India) and Swades De (Indian Institute of Technology Delhi,India)
Information Science Reference • ©2017 • 334pp • H/C (ISBN: 9781522520238) • US $190.00

Multimedia Services and Applications in Mission Critical Communication Systems
Khalid Al-Begain (University of South Wales, UK) and Ashraf Ali (The Hashemite University, Jordan & University of South Wales, UK)
Information Science Reference • ©2017 • 331pp • H/C (ISBN: 9781522521136) • US $200.00

Big Data Applications in the Telecommunications Industry
Ye Ouyang (Verizon Wirless, USA) and Mantian Hu (Chinese University of Hong Kong, China)
Information Science Reference • ©2017 • 216pp • H/C (ISBN: 9781522517504) • US $145.00

Handbook of Research on Recent Developments in Intelligent Communication Application
Siddhartha Bhattacharyya (RCC Institute of Information Technology, India) Nibaran Das (Jadavpur University, India) Debotosh Bhattacharjee (Jadavpur University, India) and Anirban Mukherjee (RCC Institute of Information Technology, India)
Information Science Reference • ©2017 • 671pp • H/C (ISBN: 9781522517856) • US $360.00

Interference Mitigation and Energy Management in 5G Heterogeneous Cellular Networks
Chungang Yang (Xidian University, China) and Jiandong Li (Xidian University, China)
Information Science Reference • ©2017 • 362pp • H/C (ISBN: 9781522517122) • US $195.00

Handbook of Research on Advanced Trends in Microwave and Communication Engineering
Ahmed El Oualkadi (Abdelmalek Essaadi University, Morocco) and Jamal Zbitou (Hassan 1st University, Morocco)
Information Science Reference • ©2017 • 716pp • H/C (ISBN: 9781522507734) • US $315.00

For an enitre list of titles in this series, please visit:
http://www.igi-global.com/book-series/advances-wireless-technologies-telecommunication-awtt/73684

www.igi-global.com

701 East Chocolate Avenue, Hershey, PA 17033, USA
Tel: 717-533-8845 x100 • Fax: 717-533-8661
E-Mail: cust@igi-global.com • www.igi-global.com

Table of Contents

Foreword

When we find word meanings vague, we have inhibitions about using them. For me, Sliding and Mode are such words. Researching the phrase, Sliding Mode in Intellectual Control and Communication, adds more meaning. Lexicographers can be ponderous, but this definition does illuminate the subject and highlight the theme of Sliding Mode. Defining the subject area, Non-Engineering Systems, becomes even richer. In general, Sliding Mode in Intellectual Control and Communication of Non-Engineering System for Humanitarian and Social Spheres is a big challenge. In the monograph Sliding Mode in Intellectual Control and Communication: Emerging Research and Opportunities, authors and professors Vardan Mkrttchian and Ekaterina Aleshina accept this challenge and invite you, dear reader, to enjoy the results of their work as set out in 9 chapters covering all aspects of such research.

Rhea Ann Ashmore
University of Montana, USA

Preface

With the expeditious rate at which research advances and new trends emerge, publishers are constantly under pressure to release these findings in the timeliest manner possible to knowledge seekers worldwide. In an effort to address this challenge and stay abreast of the latest innovations, IGI Global is pleased to announce a new initiative to produce small-scale authored monographs on niche and trending research in the areas of business, information science and technology, engineering, education, public policy and administration, medicine, and environmental science. This new product line, called Research Insights, will benefit from an abbreviated production process to ensure timeliness of publication.

When in March, 2016 IGI Global cordially invited us to help pioneer this new initiative as an author of a Research Insights book publication to be released during the forthcoming copyright year, we gladly accepted the offer. The question of the monograph theme was also on the surface as January, 2016 IGI Global published the *Handbook of Research on Estimation and Control Techniques in E-Learning Systems* (Mkrttchian et al., 2016), which presents the latest research in online learning and educational technologies for a diverse range of students and educational environments. Each of 43 chapters of the Handbook, the perspective research part contained the coverage of research advances and new trends emerging. The given monograph focuses on Sliding Mode, Intellectual Agents, and Communication. What has come out of it? You will see it for yourself, dear reader. We hope you will enjoy reading the book and, moreover, will find it useful for your future. We are sure that the reading will result in your changed perception of the outer world, in the virtual reality, at least in its part related to digital control in the Internet. The reason for this lies in the depth and panoramic scale of the change shown convincingly enough, in our view, with the author's argumentation in non-engineering systems forcibly introduced and maintained in Sliding Mode.

REFORMING SLIDING MODE FOR HUMANITARIAN AND SOCIAL SPHERES

The Sliding Mode Control finds quite a wide application in all spheres of technics and technology, in some scientific research, in the humanities, social, education and some other non-engineering areas. First of all, it is related to relay systems which, due to their simplicity, are used as numerous regulators with dual-mode control ("off/on"). Development of the theory of relay systems in the 1940s – 1950s is mainly connected with the appearance of the relay-type control actuator and vibration voltage control instruments. Later years saw the publication of a number of works on relay systems theory, offering various approaches to analyzing periodic motion and its sustainability: frequency methods, matrix method, method using z-transformation, method of analysis in state space, method of finite-difference operators. There also appeared methods of relay systems synthesis – synthesis in state space. Moreover, analysis and projecting of relay systems widely apply approximate method of harmonic linearization. In the 1960s – 1970s the concept of discontinuous control came to be associated not only with relay systems, but with the newly-emerged systems with variable structures, and also with sliding modes whose existence is possible in relay systems as well as in systems with variable structures (Mkrttchian et al., 2016).

Despite the simplicity of action principle (especially of relay systems), dynamics of discontinuous control systems is much more complicated than that of linear systems. The systems with discontinuous control allow of such effects as self-oscillations, several equilibrium positions, sliding modes and chaos. Therefore, there exist a few various theories reflecting different theoretical and practical aspects of analysis and synthesis of such systems. Meanwhile, it should be noted that research and publications pay less attention to the problem of analysis and synthesis of such systems considering their reactions to outer control and disturbing input than to the problem of autonomous movement analysis. This is especially true for the theory of sliding mode systems. The overwhelming majority of publications on this theme (with some exceptions based on the approaches different from the present work) cover only the so-called ideal sliding modes realized as control switching with the infinitely high frequency and off-line operations. With the use of methods oriented to ideal sliding modes (if the surface of the switching is formed considering all components of state vector), the analysis of the system as a tracking one proves impossible as the tracking in the above system has ideal accuracy. At the same time, it is common knowledge that in practice the accuracy of the systems is not ideal. Thus construction of the model considering the reasons for this inaccuracy and the corresponding methods of synthesis and analysis is very important for the practice of designing such systems (Mkrttchian, 2012).

Appearance of the theory of the so-called sliding modes of high and, particularly, second-order (Mkrttchian et al., 2016), the theory which has been progressing in the last decade, poses new research problems in terms of developing frequency methods of synthesis and analysis as well as in applications of this theory. Offered as means of eliminating high-frequency vibrations proper to regular sliding modes, the developed algorithms of high-order (particularly, second-order) sliding modes realization posed new questions: if they really help to eliminate high-frequency vibrations and if they offer advantages over regular sliding modes in terms of motion dynamics (Mkrttchian, 2011).

Sliding Mode in non-engineering spheres is described in a very scattered way. Though, since 2011 research and technical community has been demonstrating interest for Sliding Mode application in these spheres, particularly, in education, in research and in some other area of the humanities (Mkrttchian et al., 2016).

The goal of the monograph is new research of Sliding Mode in non-engineering systems, development of methods, technologies, software solutions, technical means and tools for increasing efficiency of intellectual control and communications with estimation of new emerging opportunities and results.

Achievement of the above goal presupposed coverage of the following aspects of the research theme:

1. The Sliding Mode Technique and Technology (SM T&T) According to Vardan Mkrttchian
2. The Sliding Mode in Intellectual Control
3. The Sliding Mode in Virtual Communications
4. The Sliding Mode in Real Communications
5. Application of Results of Emerging Research and Their Opportunities

THE RESEARCH PROJECT DESIGNING INTELLECTUAL CONTROL AND COMMUNICATION IN NON-ENGINEERING SYSTEMS

In the general flow of literature on the Internet and sliding mode control, significant segment has been taken by the works by Vardan Mkrttchian. These works focus on creation of Sliding Mode Technique and Technology (SM T&T) based on some fundamental advantages of Sliding Mode in the Internet for increasing efficiency of managing and controlling processes in non- engineering systems for Humanitarian and Social spheres. The cause of this seems clear: rapid growth of virtual universe and improvement of its "consumer qualities" in the latest years necessitates serious consideration of the Internet and digital technologies including Sliding Mode and its applications as one of the most profitable spheres for expanding the existing and realization of new business projects.

Nevertheless, the adequate choice of business-strategy is impossible without understanding of global directions of modern society evolution and the realia to determine the essence of the digital society whose outline is getting more visible for the present generation of the citizens of developed countries. Books and other publications by the Australian professor Vardan Mkrttchian, the founder and Chief Executive of the University of Control, Information Sciences and Technologies (HHH University) published by IGI Global in 2011-2017 under the general theme "Sliding Mode Technique and Technology (SM T&T)." The original title of the theme is interpreted as "Methods and technology of Sliding Mode." The interpretation seems quite proper as the issues covered in publications of IGI Global, apart from purely academic and encyclopedic ones, thoroughly deal with Control, Education, Economy, Philosophy, Psychology, Sociology, etc. it should also note that the term "Discontinuous Control and Setting Adjustment" was introduced by Mkrttchian in his Ph.D. (1979) and Dr of Sciences (1990) dissertations. The term has been recently (since mid-2015) used as Digital SM T&T. Before that, innovative solutions based, first of all, on information and computer technologies for increasing the efficiency of control, education, humanitarian and social sphere (Economy, Philosophy, Psychology, Sociology, etc.) were denoted by the term "information" or "post-industrial". The works by Mkrttchian refer to the category of publications, the authors of which try to consider the changes affecting the humanity and to determine the directions for its perspective advances in the time labeled by most scholars "information era" or "digital epoch." As a result of long-term research, Mkrttchian found out the fundamental advantages of Sliding Mode in Web for increasing efficiency of controlling processes in non-engineering systems, namely in state, municipal, social and private spheres, in continuous education, research, control and management of the whole of the humanitarian and social spheres (Economy, Philosophy, Psychology, Sociology, etc.):

- Sliding Mode in digital systems makes them unambiguous for external disturbances;
- The whole of state space is divided into two parts: a) where Sliding Mode is present; b) where Sliding Mode is absent;
- The state space is divided by the tree-dimension line, the C-border of the Sliding Mode existence;
- Sliding Mode as objective reality in the Web is transformed into virtual reality, and is controlled and managed by the summation sign of the rate of deviation from the C-border and their derivatives in time;
- All the discovered fundamental advantages of Sliding Mode in non- engineering systems are possible only with the premeditated introduction of A-space, forcible maintenance of management system trajectory on C-border.

Mkrttchian concluded that for increasing efficiency of control and managing processes in non- engineering systems, particularly, in state, municipal, social and private control, in continuous education, research, in management of the humanitarian and social spheres (Economy, Philosophy, Psychology, Sociology, etc.) key role is played by focusing on knowledge, digital form of objects presentation, production virtualization, innovative nature, integration, convergence, eliminating mediators, transformation of the "producer-consumer" relationship, dynamism, globalization (Mkrttchian & Stephanova, 2013). The practical application of the Sliding Mode in the above spheres includes the following points (Mkrttchian, 2015):

- Efficient personality (implied is a person equipped with a multimedia computer);
- Highly-productive staff (work-team interacting via computer technologies);
- Integrated enterprise (corporation with the integral inner information structure);
- Expanded enterprise (inter-corporation computer networks connecting several companies);

- Business activity in internetworking environment (global digital community – the Web).

These elements with each of the following comprises the previous ones, thus forming a new quality, according to Mkrttchian, function as links of Sliding Mode application in non- engineering spheres. This scheme logically leads to the main contents of our monograph. From the efficient personality of the multiply increased performance index the analysis passes over to the work-team whose participants reduce labor effort and greatly overcome the hampering role of the established bureaucratic structures. The integrated enterprise with information architecture oriented to the ultimate goal, gradually transforms into the expanded enterprise that is the unity of heterogeneous organizations with the entwined computer networks. This provides considerable time gains, reduction of cumulative labor effort and additional charges of all participants of the cooperation and therefore contributes to the growth of competitiveness of all participant firms. Finally, the unity of inter-corporation connections gives rise to a new virtual reality where most processes of development, production, order and distribution of the product are realized in digital form. The issues emerging in the process are solved instantly as all participants of the process have the necessary data near at hand. In his works Mkrttchian provides the main factors which collectively turn the line of sliding modes line into the basis for the new methodology and technology of intellectual control and communications. Among them, apart from the above fundamental advantages of the sliding mode, the transfer from the analog engineering to the digital one, from quasi-conductors to micro-processors, from centralized calculations to the client-server architecture, from the separate existence of the data, the text, the illustration and sound to the multimedia, from the specialized systems to the open systems, etc. (Mkrttchian et al., 2014). The extrapolation of the above trends more than anything else creates ground for the formation of digital community where sliding modes, objective virtual reality necessitate changes in controlling non- engineering processes in state, municipal, social and private spheres, continuous education, research and social sphere (Economy, Philosophy, Psychology, Sociology, etc.).

MONOGRAPH CHAPTERS

Chapter 1: The Sliding Mode Technique and Technology (SM T&T) According to Vardan Mkrttchian in Intellectual Control (IC)

This chapter covers the research into scientific and methodological framework for creating sustainable sliding modes in non-engineering systems of intellectual control, search for possibilities of self-organization of sliding modes methods and technologies enabled by accumulation of the data on their work in the process of their functioning with intellectual control.

Chapter 2: Sliding Mode in Virtual Communications

In this chapter we regard virtual communications as technology-maintained interaction realized via global networks. We will try to illustrate the sliding mode principle in virtual communications as exemplified by virtual political discourse.

Chapter 3: Sliding Mode in Real Communications

In this chapter real communication refers to the combined information process a set of hardware and software, working in the human relationship with a man capable on the basis of information, knowledge and experience, and in the presence of motivation to synthesize a new goal to make the new decision to take action and find rational ways to achieve this goal

Chapter 4: Digital Control Models of Continuous Education of Persons With Disabilities Act (IDEA) and Intelligent Agents in the Sliding Mode

In this chapter of goals of education, processes of humanization and democratization in society have led to the extension of educational institutions' rights and the tendency to regionalization of education. Therefore, the role of educational institutions in the educational space has changed.

Chapter 5: Terms of Adaptive Organization of the Educational Process of Persons With Disabilities with the Use of Open and Distance Learning Technologies (Open and Distance Learning – ODL)

In this chapter the globe has been witnessing an age in which change is an important factor and ODL is not immune to these emerging changes. Therefore, ODL should embrace the fundamental changes to survive in a rapidly advancing world. In this regard, one of the best strategies to survive and compete is to understand the administration and leadership in ODL, and identify future planning accordingly.

Chapter 6: Providing Quality Education for Persons with Disabilities Through the Implementation of Individual Educational Programs Managed by the Intelligent Agents in the Sliding Mode

In this chapter the quality of professional education for persons with disabilities is realized only in conditions considering specifics of communicative and cognitive activity of the students with different disability categories.

The absence of these conditions in universities makes it impossible for this category of students to complete the programs of higher education. As a rule, the contents of the study programs and the study schedule do not take this category of students into consideration.

Chapter 7: Regulation of Discourse in Accordance With the Speech Regulation Factors Creating Conditions for Adaptability to the Situation

In this chapter, the sliding mode technique realized in communication as speech regulation principle provides for the flexibility of discourse, namely virtual political discourse and its adaptability to the communication situation as well as to the standards. Generally speaking, virtual political communication is regulated by the standards of diplomatic discourse and censorship less than classic political communication which is connected with the anonymity of online communicators.

Chapter 8: Complex Social, Medical, Psychological, and Educational Support for People With Disability Act (IDEA)

In this chapter, we regard happens in the situation of crisis of transition from educational space to another. Meanwhile, institutions of primary, secondary and higher education reflect the acting model the psychological and pedagogical integration of persons with disabilities in society. Inclusion of persons with disabilities is regarded as a stage of their getting social adaptation and integration.

Chapter 9: Tolerance as Reflection of Sliding Mode in Psychology

In this chapter, we regard sliding mode in communication, particularly real communication, may be regarded as a technique of regulating tolerance problems. The problem of tolerance has not been fully covered despite the interest for it in various fields of science. With the accumulation of substantial empirical material, evident is the lack of generalizing research. Among scarcely covered are administration, political and professional tolerance.

REFERENCES

Mkrttchian, V. (2011). Use "hhh" technology in transformative models of online education. In G. Kurubacak & T. V. Yuzer (Eds.), *Handbook of research on transformative online education and liberation: Models for social equality* (pp. 340–351). Hershey, PA: IGI Global. doi:10.4018/978-1-60960-046-4.ch018

Mkrttchian, V. (2012). Avatar manager and student reflective conversations as the base for describing meta-communication model. In Meta-communication for reflective online conversations: Models for distance education (pp. 340–351). Hershey, PA: IGI Global. doi:10.4018/978-1-61350-071-2.ch005

Mkrttchian, V. (2015). Modelling using of Triple H-Avatar Technology in online Multi-Cloud Platform Lab. In M. Khosrow-Pour (Ed.), *Encyclopedia of Information Science and Technology* (3rd ed.; pp. 4162–4170). Hershey, PA: IGI Global. doi:10.4018/978-1-4666-5888-2.ch409

Mkrttchian, V., Kataev, M., Hwang, W., Bedi, S., & Fedotova, A. (2014). Using Plug-Avatars "hhh" Technology Education as Service-Oriented Virtual Learning Environment in Sliding Mode. In G. Eby & T. V. Yuzer (Eds.), *Emerging Priorities and Trends in Distance Education: Communication, Pedagogy, and Technology* (pp. 43–55). Hershey, PA: IGI Global. doi:10.4018/978-1-4666-5162-3.ch004

Mkrttchian, V., & Stephanova, G. (2013). Training of Avatar Moderator in Sliding Mode Control. In Project Management Approaches for Online Learning Design (pp. 175–203). Hershey, PA: IGI Global.

Acknowledgment

We have many individuals to thank for their impressive on the monograph. We would like to praise the people at IGI Global: Jan Travers, Lindsay Johnson, and Marianne Caesar, as well as Kayla Wolfe, who provided framework for the revision; they also suggested areas that could be strengthened and were invaluable in shaping the monograph. They helped make critical decisions about the transformative structure of the monograph, and provided useful feedback on stylistic issues. The final words of thanks belong to Elena Semionova and our families.

Section 1

About Theory of Sliding Mode in Intellectual Control and Communication

Chapter 1
The Sliding Mode Technique and Technology (SM T&T) According to Vardan Mkrttchian in Intellectual Control(IC)

ABSTRACT

By intellectual control we mean the total of engineering tools and software joined by the information process and working in coordination with a person (a group of people), able to synthesize goals on the basis of the data and knowledge, take decisions for action and find rational means of achieving aims. As it was mentioned in this chapter, the sliding mode data serves the ground for a new methodology and technology of intellectual control and communications. This chapter covers the research into scientific and methodological framework for creating sustainable sliding modes in non-engineering systems of intellectual control, search for possibilities of self-organization of sliding modes methods and technologies enabled by accumulation of the data on their work in the process of their functioning with intellectual control. This allows of undertaking more exact control methods which is impossible at the initial stage because of the incomplete knowledge of environment impact and the state of the system itself and, most importantly, the object of control in a non-engineering system.

DOI: 10.4018/978-1-5225-2292-8.ch001

INTRODUCTION

We offer to base any structure of intellectual control of non-engineering processes on the concept of sliding mode in animated nature which is defined as a continuous formation with the feedback information about the results of the action. Each functional system responsible for this or that adaptation effect has numerous channels along which the information from the periphery reaches the corresponding centers. The useful adaptation effect is the central point in any functional system in sliding mode as it promotes reaching the goal which in its turn serves as a system-building factor. The distinctive features of any result, even the smallest one contributing to the goal-reaching is it being produced according to the self-regulation principle irrespective of the level and complexity and possesses the same switch configuration mechanisms. In non-engineering systems these mechanisms are as follows: afforestation goal synthesis; taking decisions for action; efferent action program; action acceptor forecasting the parameters of the prospective result; return afforestation about the parameters of the result of the action; comparison of the parameters of the real result to the parameters forecast or predicted in the action acceptor. All this fits the sliding mode ways.

MAIN FOCUS IN CHAPTER

Solutions and Recommendations

We developed out the diagram of intellectual control which uses calculated and constantly controlled sliding mode in the space of non-engineering system states for adaptation, self-management and self-regulation (Figure 1).

Our system of control consists of two blocks, the first of which synthesizes goal, the second – the process of achieving the goal. The first block as the primary component features motivation (the need for something) which is combined with the information obtained with the help of the system of transducers about the state of environment and the system proper. While synthesizing the goal, the knowledge is used actively, i.e. based on the knowledge kept in the system memory, the environment and the system proper are stimulated by the trigger signals accompanied by the active evaluation of the irritants from the environment. Further the information gets into the sliding mode indicator which realizes the algorithms of action program functioning,

Figure 1. The diagram of intellectual control which uses calculated and constantly controlled sliding mode in the space of non-engineering system states for adaptation, self-management and self-regulation

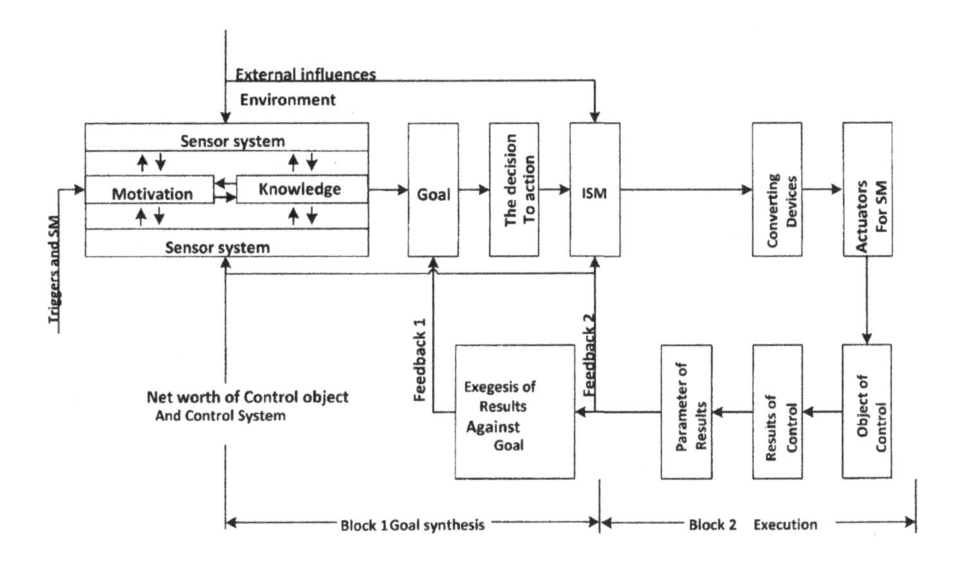

efferent irritations (control), action acceptor containing all the properties of the prospective result and serving for comparison of the predicted and the factually obtained results. Essential for the transducer is the base of knowledge. The control worked out in the transducer is realized with the help of the executive devices transforming the state of the controlled object, and the information on the result parameters by feedback link 2 gets to the transducer where the parameters of the predicted and the factually obtained results are compared. If with the worked out control the goal is reached, i.e. the difference between the result parameters satisfies the requirements, the control is functioning in sliding mode, if not, it is corrected. If it turns out that the synthesized goal is unachievable, the result parameters are interpreted in relation to the goal and the goal is corrected (feedback link 1). As a mathematical model able to describe processes going on in an intellectual system, the following correlations can be assumed,

$$T \times X \times S \xrightarrow{a_1} M \times T;$$

$$T \times M \times X \xrightarrow{a_2} C \times T;$$

$$C \times T \times X \times S \xrightarrow{a_3} R \times T;$$

$$T \times X = \{A \times T\} X \times T + \{B \times T\} U \times T;$$

$$T \times Y = \{D \times T\} X \times T; T \times R \times Y \xrightarrow{a_4} C \times T,$$

where T is the multitude of time points; X, S, M, C, R, Y – multitudes of states of the system, circumference, motivation, goal, predicted and real results correspondingly; {A}, {B} and {D} – matrices of system parameters; a_1, a_2, a_3 and a_4 – intellectual operators of control using the knowledge actual only in sliding mode.

The solution to the problem of the qualitative and quantitative description of intellectual control processes in sliding mode is connected with the necessity of defining operator's a_1, a_2, a_3 as only in this case all advantages of sliding mode are possible (Mkrttchian & Stephanova, 2013).

Our research has shown that the development of intellectual control in sliding mode should start with studying of the intuitive (associative) aspect of information processing by a human and its implementation as a new information technology – intellectual control in sliding mode. Accordingly, while coordinating logical and intuitive aspects of information processing, intellectual control in sliding mode as a new paradigm in information technology will include new functions: goal synthesis based on motivation, data about the environment and the system's state proper, integration of various, complicated with cross linkage of information containing uncertainties and getting appropriate (approximate) solution at a reasonable time; active mastering of essential information and knowledge and getting knowledge inductively; adaptation of the system proper to the consumer and changing conditions; development and performance of control for goal-achieving. As a person is able to flexibly process information due to his brain connecting the distributed information presentation, highly-parallel processing, ability for learning and self-organization, ability for information integration, in technical realization of intellectual control in sliding mode we revealed two major aspects: 1) functional - characterized by acceptability and integration of indefinite and unreliable information and ability for adaptation and learning; 2) computational - characterized by highly-parallel and distributed processing of multi-module, multi-dimensional information with a great number of links. Actually, information processing is a function, ability acquired by people in the process of evolution getting adapted to the changing conditions and effects of the environment. Though this function is multilateral, as it has been mentioned, it is possible to single out the aspects of logical and intuitive information processing. According to this, we have found out how these two aspects of information processing develop and get integrated in the computational part of the intellectual control in sliding mode in relation to information processing and working out control by non-engineering systems, by a human. The diagram of singling out two approaches in information processing in computational environment of intellectual control in sliding mode is illustrated by Figure 2.

This figure shows that there is a difference between the results of information processing by human brain and the algorithms realized in a computational environment. This difference must be minimized on account of the fact that the ability for "flexible" information processing is still a human property. Apart from this, there is a need for the synthesized control which could facilitate achieving of the goal chosen for intellectual control in a sliding mode system. Figure 2 also shows that the technology of information processing should complete or substitute a human function of its processing by means of automation and integration of logical and intuitive approaches. However, historically automation mechanisms were developing as theoretically and technologically applicable to the logical processing in traditional digital computers. Sequential processing has been therefore established today as a predominant paradigm. Still, we have studied intuitive information processing separately in such areas as image recognition and learning whose algorithms are realized in computational networks where parallel and distributed information

Figure 2. The diagram of singling out two approaches in information processing in computational environment of intellectual control in sliding mode

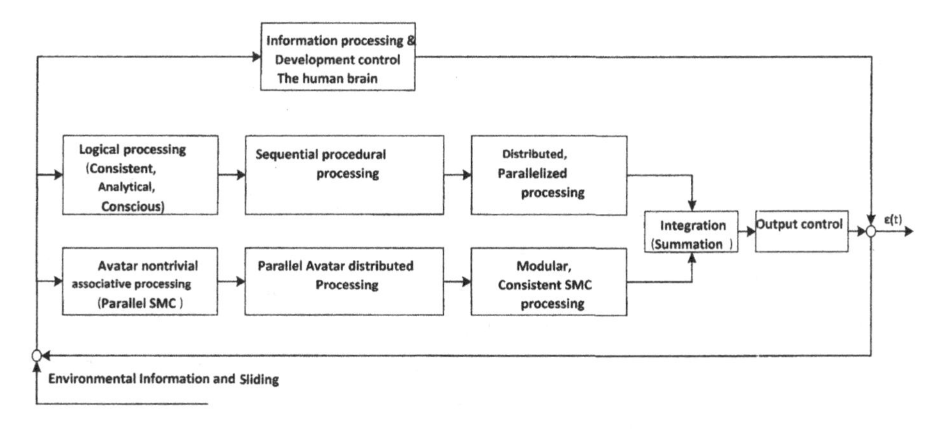

processing can be realized. Along with learning, as we have found out, such systems support various aspects of human activity accumulating various types of real-life information in the database as a result of processing and use this information for taking decisions and working out control for achieving predicted results of the whole system functioning. Studies have shown, though, that such information is wide enough and, by nature, processes modality, indefiniteness and incompleteness. Thus intellectual control in sliding mode requires realization of new functions with a various flexibility which would absorb such concepts as sustainability of functioning, the quality of processes carried out in real time, openness. The novelty of functions is based on the new theoretical approaches or algorithms suitable for intellectual control including such flexible functions as integration of symbols and images, learning and self-organization. So the technology of self-organization and effective cooperation plays an important role in this process. As flexible information processing is beyond traditional approaches, for non-engineering systems control we offer to use the concept of intellectual agent – avatar introduced by V.S. Mkrttchian, also known as avatar control technology which opens new innovative opportunities for increasing the efficiency of the process including some areas of new functions research, namely: recognition and comprehension of various types of information like pictures, speech sounds and symbolic information proper to natural languages; output and solving tasks with the help of databases allowing of direct processing and are capable of learning and self-organization; interface and modeling of interaction between a human being and the real world; control and automatic control in intellectual systems functioning in real environment (see Chapter 4). According to this, it seems possible to determine two directions of development of intellectual control in sliding mode:

- Automated intellectual systems adapted to the real environment;
- Dialog systems integrating the functions of automated systems and a human being in their interaction.

The first direction means joining intellectual systems with the real world. In that the systems should be able to autonomously comprehend and control the environment by active and adaptive interactions with the real world and are also able to undertake part of human activity in this world. Such systems are to cope with incompleteness, indefiniteness and changeability of information characteristic of the real world. The new functions of such systems include comprehension of the environment impact, planning a sequence of actions, optimal management aimed at achieving the desired outcome, elements of adaptation and self-organization.

The second direction means "integration" of the system and the human being. These should be flexible systems supporting and increasing people's intellectual activity in such areas as solving tasks and getting information by expanding data links between people and systems. In order to help people with task-solving and getting new information it will be necessary to flexibly perceive and integrate its different kinds. Here come new functions in the system: a question and an answer expressed in a real language; comprehension of intentions on the base of various information coming from people; realization of intellectual support for finding and presenting useful information in the great volume of the data kept in the databases; intellectual modeling for creating new information data and forecasting changes in the real world; integration methods for interaction of a human being and a system; a computational model of the real world, etc. For intellectual systems these new functions are to be evaluated from the viewpoint of providing such vital characteristics of intellectual systems as sustainability, openness and work in real time. Moreover, intellectual systems are already based not on one but on different key information technologies, i.e. technologies for highly-parallel computational networks (similar to transporters), optical computational systems, neuron systems and logical computational systems. These technologies can be integrated intellectual control in sliding mode to solve the tasks of the real world. The current state of intellectual systems is characterized by the situation when all spheres of large-scale human activity are affected by the ideas of intellectualization of information processes and working out control for goal-achieving. The problem of synthesizing goal as based on the data of the environment, knowledge and motivation can be solved by using the intellectual agents – avatars. Intellectual agent is an adaptive computer program modeling a certain process of the participant of control technology realization in the non-engineering sphere. This control technology is realized not only by people, but also their intellectual agents, in the virtual reality – human beings and their avatars. By doing so, the avatars build their own networks of adaptation to concrete conditions of control process in a non-engineering system. Therefor emerges a network of regulators – the people connected through their avatars with the avatars of the other participants of control process. As avatars are digital models, with them it is easier to provide sliding mode, deliberately to introduce it and to support it on the sliding edge. And integration of information processing and control algorithms in sliding mode bases capable of self-organization depending on the environment state, self-state and goal, becomes easy to realize. In its turn, this makes it possible to consider the manifestation of cooperative interaction of a great number of objects of various natures for describing such action all the way through getting a universal worldview. Intellectual control in sliding mode in a non-engineering sphere of life enables transition from "the age of closed balanced society" to the" open non-balanced society" demanding intellectual control in sliding mode as well as global standardization. Using the Internet on the basis of merging of intellectual systems with agents-avatars and information networks can put an end to geographical troubles of global cooperation (Mkrttchian et al., 2016).

CONCLUSION

Sliding mode is defined in its essence as a mathematical model of control system and invariant is space. To fully determine the spheres of its application; we made an attempt at turning to the areas of pedagogy, psychology and communication. For that, we had to transfer the sliding mode with its concrete prototype of a mathematical model to abstraction. This resulted in presenting sliding mode as abstract philosophy of adaptability in the process of continuous development of an object/ phenomenon/ system. The sliding mode can serve as basis for organizational models, techniques, methodologies that can contribute to the efficiency of solutions in the sphere of intellectual control, education and communications.

REFERENCES

Mkrttchian, V., Kataev, M., Hwang, W., Bedi, S., & Fedotova, A. (2016). Using Plug-Avatars "hhh" Technology Education as Service-Oriented Virtual Learning Environment in Sliding Mode. In Leadership and Personnel Management: Concepts, Methodologies, Tools, and Applications (pp. 890-902). Hershey, PA: IGI Global. doi:10.4018/978-1-4666-9624-2.ch039

Mkrttchian, V., & Stephanova, G. (2013). Training of Avatar Moderator in Sliding Mode Control. In Enterprise Resource Planning: Concepts, Methodologies, Tools, and Applications (pp. 1376-1405). Hershey, PA: IGI Global.

KEY TERMS AND DEFINITIONS

Chattering: It is results in chattering, an undesirable phenomenon of applying SMC, where high frequency switching is applied to the system during the sliding phase.

Construction of Equivalent Control Action: Is the system motion along the sliding surface can be interpreted as an *average* of the system's dynamics on both sides of the sliding surface.

Determination of Equivalent Control: Is the system motion along the sliding surface can be interpreted as an *average* of the system's dynamics on both sides of the sliding surface.

Discontinuous Control: It this brings the discontinuity to the control, and the whole nonlinear system.

Second – Order Sliding Mode: Is order reduction phenomenon was reviewed above and partial dynamic collapse is the reduction of order for the compensated dynamics of the SMC of the system.

Sliding Mode Control: Is control algorithm for adjustment of learning tasks.

Chapter 2 Sliding Mode in Virtual Communications

ABSTRACT

The 20th century globalization and transition to post-information society resulted in fast production and instant spread of information. Virtual communications are becoming the dominant type of communications in the institutional sphere, quite often ousting the real ones. We regard virtual communications as technology-maintained interaction realized via global networks. We will try to illustrate the sliding mode principle in virtual communications as exemplified by virtual political discourse.

INTRODUCTION

The study seems vital as politics has become an integral part of everyday reality having entered every household by means of modern media. Development of media, gadgets and the Internet has facilitated the way for political message from the addresser to the addressee and has enabled it to instantly reach the audience either target or non-target. Politics determines and forms attitudes, values and ideas. Millions of people all over the world follow the development of political conflicts, minor or major, domestic or international. Political conflict is a characteristic feature of the post-globalization era, an integral feature of modern reality (Weber, 1948; Dahrendorf, 1968; Aleshina, 2016) directly or indirectly affecting masses of people worldwide. In international relations conflicts are simultaneously a special form of political interaction

DOI: 10.4018/978-1-5225-2292-8.ch002

and a way of resolving contradictions. Generally speaking, political conflict may be regarded as a complex phenomenon and as a feature common to any political system. Political conflict may be defined as counteracting of parties (political subjects) that is expressed in certain actions directed against each other. The actions may be different in nature. Speech actions in a political conflict situation present interest for the researcher in terms of investigation of political communication in general, and political conflict communication in particular.

The study of political communication, particularly in a situation of political conflict, seems of high priority under the present circumstances of overall widespread of information and opposition in the course of information opposition. Of primary interest for a researcher is the functional and communication sphere of politics as a whole and of a political conflict in particular. According to van Dijk, reproduction of political information corresponds to reproduction of political discourse as stipulated by the ties between political actions and political processes on one side, and communication and discourse, on the other. Critical discourse analysis postulates the power being connected with control, and control over discourse means the way to its production, therefore, to its contents, style and finally, to mass consciousness (van Dijk, 2013). Another link not to be omitted is the interrelation between language and culture. Language may be regarded both as part of culture affecting its products and the instrument for creating these products (Blokh, 2013a).

Western researchers considered the issues of language and ideology prior to Soviet (Russian) ones due to substantial restrictions on Soviet scholarship. The first works on political communication describe the propaganda techniques (Lippman, 1921 ; Lasswell, 1927; Lazarsfeld, 1940; Klemperer, 1968; as cited in Aleshina, 2016). The issues of language in a state are tackled in the works by Chomsky (1988), Besancon (1984), Hahn (1997), Grenoble (2003), Duhn (1995)(as cited in Aleshina, 2016). Political discourse analysis is considered in the works by Fairclough (2003), and Chilton (2004) (as cited in Aleshina, 2016). The above researchers investigated the problems of language and language influence on mass consciousness, the phenomenon of speech manipulation in political rhetoric.

In post-Soviet Russia the relationship of language and Soviet politics came under study in the 1990-s. The issue became the subject both of linguistic and of interdisciplinary research by Kostomarov (1994), Kupina (1995),

Nikolaeva (1998) (as cited in Aleshina, 2016). The interest for political communication problems marks the fundamental works by Sheigal (2000), Wolfson, Chudinov (2006) (as cited in Aleshina, 2016). Speech manipulations in political communication were specifically studied by Baranov, Blakard (1979), Bolinger, Karasik (1992, 2002), Sheigal (2000), Pocheptsov (2000), Chernyavskaya (2006) (as cited in Aleshina, 2016).

Three main directions in researching political communication abroad were generalized and described by Budaev and Chudinov (2008) (as cited in Aleshina, 2016). Rhetorical direction represented by Anderson, Carpenter, Rikkert, Osborn, Tompson (as cited in Aleshina, 2016). Applies traditional methods of political communication analysis. Cognitive direction scholars (Lakoff, Johnson, Drulak) (as cited in Aleshina, 2016)

base their research on the cognitive approach according to which speech activity is perceived as the reflection of the worldview existing in human consciousness. The third, discursive direction is based on the discursive approach giving special attention to the situation in which the text was created and is functioning. Highlighted is the correlation between the text and the social and political conditions and cultural background of the speaker and his/ her national traditions. Within this direction, critical discourse analysis as represented by van Dijk (2013), Fairclough (2003), Ager, Maas, Wodak (2008) considers communication as a form of social practice and is aimed at studying the interaction between the language and social phenomena (Aleshina, 2016).

The essence of sliding mode for virtual communications may be disclosed in terms of communication theories, particularly, the theory of factors of speech communication regulation (Blokh, 2013) in a given situation of speech communication. We regard political conflict as determined by the specific conditions under which it develops – a situation defined as a complex of circumstances under which the counteraction of the conflicting parties is unfolding. The specifics of the research necessitate the definition of political discourse as central to communication analysis. Following the idea of discourse being a more general concept than text (Blokh 2013a), we define political discourse as a text determined by the theme of expression and asserting interests of political subjects considered in a situation of political communication. Political discourse incorporates a wide range of genres of political communication, both in classic and online forms.

In the process of illustrating how sliding mode functions in discourse, we regard the text-discourse as consisting of dictemes (Blokh, 2000). Dicteme is defined as a basic unit of language and a unit of text schematization which provides for the sequence of speech. The dicteme realizes four principal functions of speech: nomination, predication, schematization and stylization. According to M.Y. Blokh who introduced this concept in linguistics, speech acts are realized in dictemes of the text (Blokh 2013b). This view contradicts the theory of speech acts developed by Austin who stated that speech acts are realized in assertions (sentences) (as cited in Aleshina, 2016). . Dictemes informemes as minimums of discourse (text) serve to actualize the speech acts. Accordingly, communication may be described as exchange of dictemes.

In our view, dictemes as correlating with speech acts (both illocutionary and elocutionary) may be nominated according to the types of speech acts. For example, a text of conflict situation may consist of dictemes-accusations, dictemes-reproaches, dictemes-verdicts and the like. Nomination corresponds to different types of atonality/ aggression speech acts.

According to M.Y. Blokh, the unfolding dicteme corresponds to the exposition of the information aspect of the text. The dicteme information complex realized in the course of the speech act comprises several types of information rubrics reflecting important features of cognition and communication. The first rubric is communication and guideline information defining the type of cooperation between the speaker and the listener in terms of speech and activity; the second rubric includes factual information of general type reflecting the situation. The third rubric features factual information of special type conveying ethnic, social and cultural, terminological realities. The fourth rubric comprises intellectual information reflecting the movement of the cognizing thought of the speaker. The fifth rubric comprises the emotive information connected with emotional expression. The sixth rubric consists of structural information that conveys the structural properties of the text that typologically mark it. The seventh rubric is register information reflecting differences between neutral, literary and colloquial variations of speech. The eighth rubric comprises social and style information corresponding to the text functional style. The ninth rubric features dialect information reflecting territorial and ethnic properties of the text. The tenth rubric is impressive information implementing the connotation of target influence on the listener. The eleventh rubric contains esthetic information that builds the aspect of artistic way of expression. Each of these rubrics is linked to the special content type of dicteme. The generalized classification offers four main types of dicteme information: factual, intellectual, emotive and expressive (Blokh, 2000).

Turning to the theory of factors of speech communication regulation, we can observe the correspondence between those and the structure of a conflict situation (Aleshina, 2014). We follow the classic concept of conflict structure offered by Weber (as cited in Aleshina, 2016).

According to him, the conflict structure as an ideal static phenomenon includes the following elements: two or more conflict sides (subjects of conflict, conflicting parties); the object of conflict; indirect conflict sides (organizers of conflict, provokers); third side (mediators, referees); social environment (Kozyrev, 2008).

Blokh offers seven factors of regulation of speech communication (Blokh, 2013b). The first factor is the target content of speech (utterance), defining what is being said and what it is being said for. These two points build the communicative frame of the meaning of the utterance. The second factor is the personal status of the speaker which includes characteristics contributing to the formation of the speaker's linguistic identity. Personal characteristics include temperament, ethos, moral qualities, level of education, occupation, social status, intellectual abilities. The third factor is the personal status of the listener. The listener's characteristics correspond to those of the speaker. Meanwhile similar features realize different regulation consequences in different communicative positions (that of the listener and that of the speaker). The fourth factor of regulation of speech communication is the presence or absence of the persons who hear the speech but are not involved in communication. To this category refer those who are unwillingly listening or those eavesdropping. The fifth factor is the properties of the communication link. The sixth factor, pre-supposition, is defined as the assumption of the speaker about the background knowledge of the listener. The image of the listener who the speaker addresses may be different from the real personality of the listener. The seventh factor is the post-supposition, the assumption of the listener about the personality of the speaker. Non-productive (post-suppositional unjustified) communication results from the discord in the ideas of listener and speaker of each other. These may lead to awkward situations with discursive lapses known as gaffes. Political gaffes may cost a politician his/ her career.

MAIN FOCUS IN CHAPTER

Solutions and Recommendations

In its general sense, sliding mode in communication, including virtual communication, may be regarded as a technique of regulating discourse in accordance with the above speech regulation factors. Thus, communication gets adapted to the situation (in our case – the situation of political conflict) in the framework of personal, cultural, situational regulators, strives to perform its functions and achieve goals. We singled out the following major goals of political discourse: information, persuasion, invocation, penance. These functions correspond to the genres of political discourse (Aleshina, 2016) defined on the principle of goal-setting of the utterance.

The target content of speech (utterance) as the major factor of regulation of speech communication determines the outline of communication. Current political public communication is not only delivered by the speakers in a live mode but also broadcast and published online. That explains the fact that the target content of speech (utterance) in online political communication does not offer many differences from classic communication in terms of information and content expression. Political discourse is mainly built on certain diplomatic regulations that maintain the rules generally observed in public communication. The utterances-dictemes of the text either real or virtual, of political conflict communication are marked by the atonality connected with the controversies of domineering ambitions. This feature characterizes speech acts of political conflict communication (speech acts of accusation, demand, mockery, etc.). The information (factual, intellectual, emotive, impressive) actualized the dictemes determines the degree of atonality in the speech act (Blokh& Aleshina,, 2015).

The variations lie in the less official register of political communication. Online political discussions and blogs, especially those with anonymous posts, offer wider opportunities for making the utterance as impressive and emotive as possible which could be hard to achieve with restrictions and censorship. In virtual political communication we can find a wide range of texts referring to various registers of speech, from intensely formal to intensely informal (Aleshina 2016).

The second factor of regulation of speech communication, personal characteristics of the speaker in a political conflict situation is traditionally determined by political communication norms. We also argue that the style of behavior and communication of a political conflict participant directly correlates with the style of conflict behavior modes described by Thomas and Kilmann (Aleshina, 2014). There are two main strategies of conflict behavior – the strategy of partnership and that of assertiveness. Within these strategies the researchers single out several tactics of conflict behavior (Morozov, 2002):

Avoidance characterized as unwillingness to see and recognize the differences, denying the conflict itself; concession marked by the aspiration to establish and improve relations by means of smoothening contradictions; confrontation (competition, rivalry) is linked to the wish to stand one's ground by means of open confrontation; compromise is characterized by the aspiration to come to terms by means of negotiations; cooperation is connected with the search for solutions acceptable for the conflict parties.

Despite the mentioned diplomatic rules and regulations, political leaders may stick to different strategies and tactics in a situation of political conflict. These strategies and tactics are similarly expressed both in classic and online official political communication. The second factor of regulation of speech communication is closely connected with the first one as the strategies can be traced in the content of the utterance.

The third factor of regulation of speech communication is the personal status of the listener. In terms of current political communication we should also add the factor of the reader to this item: the personal status of the listener/ reader. The success of communication depends much on his/ her personal background including education, social status and some personal psychological characteristics such as potential conflict behavior strategies. The political text may be incorrectly perceived due to the addressee's negative experiences connected with the issue in question or the lack of expertise in it. Online communication allows for the listener's/ readers immediate reaction in terms of expressing approval/ disapproval, support or objections (see, for example, the Politics section of the New York Times or Debate Politics forums on debatepolitics. com). Instant messaging facilitates feedback to political moves as compared to the 20th century correspondence of the listeners/ readers with editorial boards of the media and headquarters of political figures.

The fourth factor of regulation of speech communication is the presence or absence of the persons who hear/ read the speech but are not involved in communication. This factor is closely connected with the fifth one – the properties of the communication link.

Classic political communication tends to follow the diplomatic canons especially if it is public discourse. The violations of norms are less likely to be seen by the public on TV, heard over the radio. On the contrary, online political communication is marked by involvement of big audiences whose most parts do not participate in the process as speakers or writers but are readers. Online discussions participants are well aware of the fact that they are being 'watched', their words are being read, answered and reposted. The repost phenomena typical of online communication contributes to the wide spread of political information and global processes of influencing mass consciousness.

The sixth factor, pre-supposition deals with the image the speaker/ writer has of the listener/ reader. This factor is closely connected with the seventh factor, the post-supposition, the assumption of the listener / reader about the personality of the speaker/ writer. Non-productive (post-suppositional unjustified) communication results from the discord in the ideas of communicators have of each other. These may lead to awkward situations with discursive lapses known as gaffes (Aleshina, 2015).

To sum up, it should be noted that virtual political communication, similarly, to classic communication, is connected directly with assertion of own political interests, the main motive being need for domination. The theory of factors of speech regulation and the theory of dicteme applied prove this similarity with the corresponding similarities in the target content of the speech, the communicators' properties and their suppositions and assumptions about each other. Virtual political communication is less regulated by the standards of diplomatic discourse. The uncensored online discussions provide more opportunities for realization of conflict and information war strategies. Instant feedback allows the online communicator to correct the non-productive political communication.

CONCLUSION

Virtual communication seems to be another sphere of sliding mode application and functioning. Regulations of discourse in accordance with the speech regulation factors create conditions for adaptability to the situation. Virtual political communication, similarly to classic communication, is connected directly with assertion of own political interests, the main motive being the need for domination. The theory of factors of speech regulation and the theory of dicteme applied to the material prove this similarity with the corresponding similarities in the target content of the speech, the communicators' properties and their suppositions and assumptions about each other.

REFERENCES

Aleshina, E. (2016). Structural, Information, and Regulation Aspects of Political Online and Classic Communication. In V. Mkrttchian, A. Bershadsky, A. Bozhday, M. Kataev, & S. Kataev (Eds.), *Handbook of research on estimation and control techniques in e-learning systems* (pp. 329–341). Hershey, PA: IGI Global. doi:10.4018/978-1-4666-9489-7.ch023

Aleshina, E. Y. (2014). Regulation factors of speech communication in a political conflict situation (on the basis of the English language). *Political Linguistics*, *2*(48), 108–113.

Aleshina, E. Y. (2015). Discursive gaffes in political conflict communication (based on the English language material). Language and culture in the era of globalization. In *Proceedings of the Second International Conference*. Saint Petersburg, Russia: Saint Petersburg State University of Economics.

Besançon, A. (1984). *Russian past and Soviet present*. London: Overseas Publ. Interchange.

Blakard, R. M. (1979). Language as a Means of Social Power. In *Pragmalinguistics* (pp. 131–169). The Hague: Mouton.

Blokh, M., & Aleshina, E. (2015). Discursive expression of the strategy of intimidation in a political text of a conflict situation (based on the English language). *Political Linguistics, 2*(52).

Blokh, M. Y. (2000). Dicteme in the level structure of the language. *Issues of Linguistics*, *4*, 56–67.

Blokh, M. Y. (2013a). Discourse and systemic linguistics. *Language, Culture, Speech Communication*, *1*, 5–9.

Blokh, M. Y. (2013b). Language, culture and the problem of regulating speech communication. *Language, Culture, Speech Communication*, *2*, 5–9.

Budaev, E. V., & Chudinov, A. P. (2008). Foreign political linguistics. Moscow: Flinta Nauka.

Chernyavskaya, V. E. (2006). *The discourse of power and the power of discourse. The issues of speech influence*. Moscow: Flinta Nauka.

Chilton, P. (2004). *Analysing political discourse: Theory and practice*. London: Routledge.

Chomsky, N. (1988). *Language and Politics*. Montreal: Black Rose Books.

Chudinov, A. P. (2006). *Political linguistics*. Moscow: Flinta Nauka.

Duhn, J. (1995). The transformation of Russian from a language of the Soviet type to a language of the Western type. In *Selected Papers from the 5th World Congress of Central and East European Studies Language and Society in Post-Communist Europe*. Warsaw: Academic Press.

Fairclough, N. (2003). *Analysing discourse*. London: Routledge.

Fairclough, N. (2003). *Analysing discourse*. London: Routlege.

Grenoble, L. (2003). *Language policy in the Soviet Union*. Dordrecht: Kluwer Academic Publishers.

Hahn, L. (1998). *Political communication: rhetoric, government and citizens*. State College, PA: Penn State University.

Klemperer, V. (1968). *LTI. Notizbuch eines Philologen*. Leipzig: Reclam.

Kostomarov, V. G. (1994). *The language taste of the epoch*. Moscow: Pedagogics-Press.

Kozyrev, G. I. (2008). *Political conflictology*. Moscow: Forum Infra-M.

Kupina, I. A. (1995). *Totalitarian language: vocabulary and speech reactions*. Ekaterinburg: ZUUNC.

Lasswell, H. D. (1927). *Propaganda technique in the world war*. New York: Academic Press.

Lazarsfeld, P. (1940). *Radio and the Printed Page: An Introduction to the Study of Radio and Its Role in the Communication of Ideas*. New York: Duell, Sloan, and Pearce.

Morozov, A. V. (Ed.). (2002). *Social conflictology*. Moscow: Academy.

Nikolaeva, T. M. (1998). Linguistic demagogy. In Pragmatics and intensionality issues (pp. 66-76). Moscow: Academic Press.

Pocheptsov, G. G. (2000). *Information wars*. Moscow: Refl-Book, Vakler.

Sheigal, E. I. (2000). *Semiotics of political discourse* (PhD Dissertation). Volgograd.

van Dijk, T. (2013). *Discourse and power*. Moscow: LIBROKOM.

ADDITIONAL READING

Aleshina, E. Y. (2012). English political public speech of a conflict situation. *Bulletin of the Moscow State Regional University. Series. Linguistics*, *6*, 88–96.

Aleshina, E. Y. (2014). Communication of political conflicts as the object of linguistic research (based on the English language material). In *Proceedings of Conference on the Problems of Education in the Humanities*. Penza: Penza State University.

Aleshina, E. Y. (2014). Target content of the utterance as a factor of regulation of speech communication in a situation of political conflict (based on the English language material). In Proceedings of All-Russia conference "2nd Avdeev Readings". Penza: Penza State University.

Blokh, M. Y., & Aleshina, E. Y. (2015). Political text at seven stages of its dynamics. *Bulletin of the Moscow State Regional University. Series. Linguistics*, *2*, 6–11.

Briggs, C. L. (Ed.). (1996). *Disorderly discourse. Narrative, conflict and inequality. New York. Oxford: Oxford University Press. Fairclough, N. (1995). Critical discourse analysis: the critical study of language*. London, New York: Longman.

Greene, R. (2007). *The 33 strategies of war*. Joost Elffers Books.

Medhurst, M. J. (1997). *Cold war rhetoric: strategy, metaphor and ideology*. Michigan State University Press.

Shiffrin, D., Tannen, D., & Hamilton, H. E. (Eds.). (2001). *The Handbook of discourse analysis*. Blackwell publishers.

Weiss, G., & Wodak, R. (2003). *Critical discourse analysis. Theory and Interdisciplinarity*. Palgrave Macmillan.

Wodak, R., & Kryzyzanowski, M. (Eds.). (2008). *Qualitative discourse analysis in the social sciences*. Palgrave Macmillan.

KEY TERMS AND DEFINITIONS

Agonality: Is nomination corresponding to different types of aggression speech acts.

Dicteme: Is a unit of thematization in a text/discourse that performs a number of functions.

Chapter 3
Sliding Mode in Real Communications

ABSTRACT

In this chapter, real communication refers to the combined information process a set of hardware and software, working in the human relationship with a man capable on the basis of information, knowledge and experience, and in the presence of motivation to synthesize a new goal to make the new decision to take action and find rational ways to achieve this goal. As it was mentioned in Chapter 1, the way of sliding modes forms the basis for new methods and technology of communications providing for adaptability and at the same time invariability of communications.

INTRODUCTION

In the last decade the term "tolerance" has established itself in academic literature (Lat. tolerantia – patience, condescension). The phenomenon of tolerance is widely researched by scholars from different science branches. This term has appeared and is actively discussed in cultural studies, sociology, political science, economics, psychology, history, pedagogy. The key idea of the research is it being the necessary framework for all productive and harmonious relationship between people.

Despite the frequent use of the term, there is still no unanimity in its understanding. There is a wide range of divergence in the views of tolerance and its limits. This is connected with different approaches in its studies and the

DOI: 10.4018/978-1-5225-2292-8.ch003

complexity of the phenomenon proper. According to the context of studies, tolerance is filled with a specific sense. Some scholars write about tolerant relationship and attitudes, others – about tolerance as a personality's trait, still others – about the skills of tolerant behavior. There exist approaches to tolerance as a core value, a form of social interaction, a culture of conversation, a significant personality quality of the specialists working with people – doctors, psychologists, politicians, managers, educators.

Numerous studies focus of difference aspects of tolerance as its essential features, connecting this concept with different factors and determiners. The word tolerance is not quite common in Russian reality. We considered the semantics of this word in different languages using foreign language dictionaries.

In English dictionaries tolerance is understood as religious toleration, assumption of religious freedom, tolerance is a skill of tolerantly accept another person's (different from own) opinions, convictions, traditions, characters, to demonstrate tolerance, to bear, to tolerate (Farber,2016).

In French tolerance supposes respect another person's freedom, his/ her way of thinking, behavior, political and religious views.

In Chinese is being tolerant means allowing of and demonstrating goodness towards others.

In Persian tolerance is patience, endurance and readiness for reconciliation.

In Arabic – forgiveness, condescension, compassion, benevolence, patience, goodwill for others.

General information about tolerance provided by the dictionaries can be summed up to the effect that tolerance is toleration and condescension, admittance of something alien, the object of toleration being "others' opinion, faith and behavior".

In the Russian language there are two words with a similar meaning – "толерантность" and "терпимость". As the analysis has shown, these concepts are interconnected and they coexist but they are not full equivalents. This can be mainly explained by the fact that the understanding of tolerance has not yet been completely formed for a Russian speaker.

The concept of "терпимость" is widely used in medicine and biology as connected with the adaptation processes in the body. In this case the axiological meaning of tolerance is determined by the body's reaction to the environment. This reaction is expressed in the rising sensibility of the body, its cells and tissues to the effect produced by a substance and contribute to maintaining homeostasis (relatively dynamic consistency of composition and qualities of interior environment and sustainability of the main physiological functions of the body). Consequently, tolerance provides for the homeostatic balancing of an individual's relationship with the environment.

Tolerance is connected with adaptation processes not only in the sphere of biology, but also in the social sphere. This clearly demonstrates the definition of tolerance in psychological dictionaries: "Tolerance is the absence or weakening of reaction to an unfavorable factor as a result of lowering sensibility to its effect. It is manifested in self-restraint, composure, the ability to bear unfavorable effects for a long time without decreasing of adaptation abilities".

K.A. Albukhanova-Slavskaya considering historic and modern features of Russian mentality notes that the Christian concept of compassion – patience plays a key role in the Russian national character. Though tolerance connected with accepting suffering, fills this concept with the content which is quite far from tolerance (Rogers, 2015).

"Толерантность", unlike "терпимость", expresses rather an active than passive position of an individual towards another individual.

MAIN FOCUS IN CHAPTER

Solutions and Recommendations

Our research has shown that the concept of the first word is much wider than that of the second due to the rational attitude to reality implied by it. The meaning of the word "толерантность" actualizes a psychological aspect which is possible to refer to the human's higher moral qualities (benevolence, kindness, heartiness, sensitivity, emotional generosity). We found out that tolerance is a conscious position, a form of civilized perception of reality, of civilized attitude to all "alien", "foreign", to dissenting views. Tolerance supposes tolerating all ideas and convictions of the people, though it does not suppose that an individual has to share all these opinions and beliefs. So, tolerance ("толерантность") in its core does not coincide with toleration ("терпимость"). From the diversity of tolerance manifestations we see that as a personal trait tolerance is formed under the influence of numerous factors and variables. They determine a personality's general positive attitude based on the ability of an individual to establish positive relationship with other people and shape an image of self. Tolerant personality is considered as opposed to the intolerant one characterized by the specific cognitive and individual features. The semantics of the word "tolerance" comprises a clearly defined idea of measure and limit up to which it seems possible to tolerate another one if his/ her actions produce bewilderment or resistance. Social and psychological limits of tolerance result from the socialization-formed inner moral and ethical (politeness, morality) and outer legislative-juridical (law-obedience, legal literacy) normative limits of a person as an individual and a citizen. It is not about understanding and accepting another one as he/ she is, with all his/ her drawbacks, it is rather about the strict maintenance of the access behind which tolerance transforms into the opposite. According to the variable flexibility of tolerance limits, determined is the tolerance measure which has its own specific features in various cultures, historic periods and certain behavioral situations. The limits of tolerance are defined by a number of physical, psychological and socio-cultural possibilities of allowing "another one's" freedom. Our research has shown that tolerance should not be equaled to indifference, narrowed down to the necessity of overcoming the feeling of disliking another one. It supposes an interested attitude to another one, a wish to feel his/ her "difference". Such understanding provides for expanding own social experience and enriching it with the new cultural background. Generalizing various approaches to understanding the essence of tolerance, it can be stated that tolerance is a social and psychological phenomenon manifesting itself through the way of interacting with the outer world. Tolerance is also expressed in the aspiration to reach mutual understanding and concord in the process of communication and interaction by means of cooperation and dialogue, in a person's ability to accept another one with all its differences. We offer to define tolerance as an individual's integral characteristic shaping its ability to actively and positively interact with the outer world in problem situations. It can be regarded as an important component of the life stance of a person possessing his/ her own values and interests ready to be protected, if necessary, and also respecting others' stances and interests. We determined the following directions of understanding the phenomenon of tolerance in modern science:

- Tolerance on the level of psychophysical toleration of negative effects;
- Tolerance as a personal characteristic which manifests itself in communication – tolerant attitude to another one's individual features; accepting another one's uniqueness.

From the viewpoint of the psychophysical approach tolerance is a person's endurance or resistance against different influences which he/ she can regard as "harmful". The lowering of sensitivity to the influence of such negative factors can be connected with the absence or weakening of the reaction to them. As an personality's psychic stability can be defined as an individual's ability to resist outer effects upsetting the person's neuron-psychic balance and independently, rather fast, to return to the balanced psychic state. Herewith there occur no violations of the person's adaptive abilities, the body functions by means of increasing sensibility thresholds without damage to health. The body's cooperation and the effects on it play an important role here. Within this approach the term "tolerance" is considered as connected with a problem of a person's ability to survive negative effects (Shirom, 2015).

Our research has demonstrated that among the psycho-physiological parameters providing for the efficiency of activity and the specifics of the tolerant behavior in strenuous conditions, the force of the excitation process, flexibility and sustainability of neural processes matter much. Thus, we state that the undertaken research into the essence of tolerance as a personality's psychophysical balance, in real communications, with application of sliding modes and intellectual agents-avatars are quite unique and ensure invariability and safety when it comes to studying human reactions in problem, crisis or extreme situations. To these refer the situations posing danger for life (technology-related and natural catastrophes, natural disasters, participation in military operations, terrorist acts), and the situations connected with professional activity demanding from a person neuro-psychic stability, ability to react quickly and confidently to changing conditions and take decisions. In these cases to the foreground comes a person's psycho-physiological resistance to negative effects and extreme situations which can serve a good ground for forming a tolerant personality.

Simultaneously, in our research social and psychological tolerance is treated as a characteristic of the peoples' attitude to people, demonstrating the degree of tolerating unpleasant or inappropriate, in their opinion, psychic states, qualities and the degree of tolerating unpleasant or inappropriate psychic states, qualities or actions by interaction partners. Tolerance manifestations are determined by an individual's not perceiving differences between self and other people or not having negative emotions about the differences. So tolerance means active acknowledgement of another opinion. At that the individual does not agree with another, different view, but recognizes its right for existence. Tolerance is rather active than passive. It does not mean refusing from own views and convictions, testifies to the dialogue communicators' openness and their "inter-permeability". Positive idea of tolerance is reached through clearing out the manifestations of its opposite's – intolerance or intolerability which is based on the assumption that your system of views, your way of life, the group where you belong are superior to the rest. Intolerability is rather conservative; it tries to suppress everything that does not fit the existing standards. The range of results of intolerability is quite wide: impoliteness, neglect to others and deliberate humiliation of the people. Social and psychological sustainability supposes sustainability to the world diversity, to ethnic, cultural, social and worldview differences. On this level it is expressed through the system of social attitudes and value orientations. This system is able to maintain neuro-psychic balance in different situations of life. The aspects and manifestations of this quality include empathy, altruism, trust, cooperation, aspiration for dialogue and others. Tolerant rates are characterized by preferring the values of kindness, independence, universalism, internality of control locus, the absence of expressed availability, etc.

We found out the basic tolerance criteria within the above approach are the following:

- Subjective activity whose indicators are social and personal self-identification, optimistic view of life, initiative, active life attitude of a person;
- Personality sustainability demonstrated in emotional stability, independence, responsibility, self-confidence and reflexivity;
- Variability of thinking – the ability to solve problems, absence of stereotypes in perception, flexible and critical thinking.
- Value-conscious, respective attitude to a person, its main indicator being the presence of emphatic abilities and benevolence;
- Aspiration for the personality's self-development and self-actualization of inner resources;
- Organization of volitional self-control.

Thus, it was determined that tolerant people are characterized by being oriented to the relationship with others and at the same time, relative independence. They are inclined to search for and find balance between own interests and other people's interests. So there is a certain sliding mode borderline along which real communications are built. Generalizing various approaches to understanding the essence of tolerance, we found out that tolerance is a social and psychological phenomenon manifesting itself through the way of interaction with the outer world. Simultaneously, we formulated the following concepts that served the basis for describing practical application of our research on real communications in sliding mode presented in Chapter 9.

CONCLUSION

Summing up the material of Chapter 3,

- We highlight the fact that the circle of problems related to tolerance as basic tool of real communications in sliding mode, is much wider than those considered above. Some have been just indicated, solving the others is on the stage of further studies within the framework of definite empirical research.
- For instance further consideration is required for the contents of several aspects of tolerance (psycho-physiological, ethnic, communication, etc.) characterizing the process of people's successful interaction.
- There arises a necessity to study the problems linked to defining tolerance specificity in various kinds of communication (family, professional, international, etc.) at different stages of life and professional realization. The study of the concepts under consideration applicable to solving the problems of a personality's successful realization is perspective research direction and will be presented from the practical viewpoint in Chapter 9.

REFERENCES

Farber, B. (2016). Understanding and treating burnout in a changing culture. *Psychotherapy in Practice*, *56*(5), 589–594. PMID:10852146

Rogers, C. (2015). Empatic: An unappreciated way of being. *The Counseling Psychologist*, *5*(2), 2–10. doi:10.1177/001100007500500202

Shirom, A. (2015). Reflections on the study of burnout. *Work and Stress*, *19*(3), 263–270. doi:10.1080/02678370500376649

KEY TERMS AND DEFINITIONS

Tolerance: Is an integral personal quality of an individual providing for its ability to actively and positively interact with the out world in problem situations. It functions as an important component of the life attitudes of a person possessing of own values and interests are able to protect them, if necessary, at the same time treating others' positions and values with respect.

Tolerant Personality: Is a personality possessing of a certain set of qualities providing for its constructive professional development characterized by skills of tolerant interaction. Tolerant personality is characterized by the following manifestations: value preference of kindness, independence, responsibility, absence of marked value, empathy, cooperation, activity, prevalence of active strategies of coping with difficulties, the ability to flexibly reconstruct the system of self-regulation as prompted by changing outer and inner conditions, high degree of reflexivity, manifestation of the internal control locus, optimism, emotional stability, ability to control impulsive behavior, responsibility, appropriateness of self-assessment and positive self-perception, extraversion, openness to the new experience.

Section 2

Application of Results of Emerging Research and Their Opportunities

Chapter 4
Digital Control Models of Continuous Education of Persons with Disabilities Act (IDEA) and Agents in Sliding Mode

ABSTRACT

Currently higher professional education is defined as a sphere of nationwide strategic interests of the state whose priorities contain systemic approaches and solutions, values of world and national culture, humanist morals, civic consciousness, worldviews and methodological solutions targeted at training new generations of specialists capable of creative activity and professional responsibility. In this chapter of goals of education, processes of humanization and democratization in society have led to the extension of educational institutions' rights and the tendency to regionalization of education. Therefore, the role of educational institutions in the educational space has changed.

DOI: 10.4018/978-1-5225-2292-8.ch004

INTRODUCTION

We consider the problem of building up a whole educational space in the context of implementing the system of continuous professional education in an integrated educational institution. The pedagogical approaches to the realization of integration processes in the system of continuous education are based on the philosophic ideas considering the man as the highest value in society. So, the aim of societal development is a person's continues moral and spiritual, personal and professional perfection (Mkrttchian et al., 2016).

The defining condition of working out digital models for the system of continues professional development of persons with health limitations is based on our concept of continues professional education which is considered as an instrument of economic policy directed at increasing competitive capacity, full-time employment of the population and maintaining employees' professional mobility as connected with implementing new technologies. It is based on the following principles:

1. Basic property (basic education) realized through getting a ceratin educational start, that is basic training is regarded as a "matriculation certificate";
2. Multiple-level system presented by a number of levels and stages of education;
3. Diversification which supposes extension of the activity types of the education system as well as acquiring new forms and functions previously absent in the system;
4. Complementarity of basic and postgraduate education referring to the vector of professional skills and progression of a person in educational space – the idea is backed up by the fact that in the system of continues education a person has to continue his/ her education for life;
5. Flexibility of educational syllabi providing for a person's orientation in Educational space, professional re-orientation, for the possibility of changing activity sphere at a certain stage of life and at a certain level of education, or getting parallel education in two or more spheres;
6. Succession of educational programs necessary for a trainee, a student, a specialist to freely move in educational space;
7. Integration of educational structures viewed as integration of subsystems of education, turning of professional educational institutions into multi-specialized, multi-level, multi-stage educational institutions;

8. Flexibility of organization forms revealed through a person's free movement in educational space which ensures not only the diversity of education forms, but also their flexibility and variability.

Creating the system of quality of professional education is a very important direction of realization of models. We consider the problem of quality from three perspectives: the quality of specialists' technological training; the quality of their economic and market training; the quality of basic qualifications formed in them. From this viewpoint we state the necessity of quality education being aimed at marketing values completely different from the traditional training typical of former planned economy. With that, a graduate will possess basic qualifications including the training components off-professional or over-professional character (information technologies skills, foreign language skills, advertising marketing skills, etc.). In present-day conditions, such training is vital for a specialist in accordance with the "formula" of successful employment at the labor market: higher education (any), knowledge of a foreign language, computer skills.

MAIN FOCUS IN CHAPTER

Solutions and Recommendations

Our research has found out the following major directions of further improvement and modernization of professional education (Mkrttchian & Belyanina, 2016):

1. Realization of advanced professional education: the level of specialists' general and professional education, development of their professional qualities and personality as a whole should be ahead of development of industry and its technologies.
2. Innovative higher educational institutions integrating primary, secondary and higher professional education in one institution and legal fixation of their status. There are two possible variants of creating such integrated educational institutions:
    a. Based on a new type of college;
    b. Based on a higher education institution by means of integrating primary and secondary professional education.

The graduate of such an integrated educational institution has knowledge and skills of an integrated specialist according to the state standard is a practice-oriented specialist qualified as a worker and possessing fundamental knowledge and skills of a modern employee able to use information technologies and the knowledge of at least one foreign language on the professional basis in his/ her practical activity.

3. Personal orientation, differentiation and individualization of education backed by educational standards based on diversity and variability of educational institutions successfully implementing the above concept of continues professional education.
4. Creating information space in every educational institution and wide use of state-of-the-art pedagogical technologies in education process together with working out and accepting the corresponding standards for their realization.
5. Creating educational, study-research-industrial and sociocultural complexes, resource centers, training and consultation centers, development of distance learning, their regulation basis.
6. Introduction of subject-oriented instruction in senior grades and integration of high school into educational institutions of primary, secondary and higher professional education.

Modern system of continues professional education guarantees building up the efficient market of education services in accordance with the economy's demands for qualified staff as based on the constant monitoring of labour market. Continued education provides every person an institutional opportunity to form an individual education trajectory and get the professional training necessary for his/ her further professional, career and personal development. We are sure that researching into the ways of the system's evolution can contribute to the growth of education's adaptability to external demands including those of labor market.

Maintenance of the growing demands for constant professional development or re-training, in our view, necessitates creation of the infrastructure of the access to continues professional education during the whole of professional activity.

Defined below are the basic elements of this infrastructure:

- Programs of professional training, re-training and development based on module principles;

- Social and professional organizations whose activity is directed at formation of labor market-stated qualification demands for the specialists' level of training, search and choice of information technologies and also the evaluation (accreditation) of the syllabi quality;
- Common system of credits based on the modern information infrastructure of control, storage and accumulation of date on education and training obtained by a person in various educational organizations;
- Nationwide system of evaluating quality of education independent of organizations, aimed at providing unity of educational space through providing citizens with an opportunity of objective control of the knowledge and skills they got.

Our research has shown that formation of the efficient market of education services, ensuring competitiveness of national education and improving its quality requires reconsideration of the list of organizations entitled to providing educational services of continuous professional education. This list should include major industrial, commercial or other organizations possessing resources for realization of various educational programs in the framework of intercompany tuition. This necessitates improving the quality of state educational institutions activity which seems possible only through transition from managing educational institutions to managing educational programs. Therefore, such administrative functions as control, financing and assessment of activity quality change their nature and are realized in reference to educational programs. This requires creating crucially new mechanisms of evaluation and accreditation of educational programs which, in its turn, will require building up organization and legal conditions for development of social and professional organizations whose members will include representatives of professional associations, education community and employer associations. We have found out that the main task of such organizations is outlining the requirements for the level of necessary professional qualification, content and technologies and training, and also for the employees' competences. The above social and professional organizations are already able to ensure efficient control of the quality of educational process and adequacy of the trained staff to the dynamic perspective demands of the labor market.

The involvement of the employers into participation in the monitoring of the labor market and formation of the list of training specialties seems to be an innovative solution and greatly improves the quality of continues professional education. Creation of the common system of credits will ensure general recognition of the education outcomes, transition to managing education programs. Spreading of the module principle of building programs will guarantee an institutional opportunity for studying in various educational institutions within the framework of the same educational trajectory of a person. With that, the options of educational programs have extended, staff training have acquired a target character and have become more efficient. We propose creating an independent nationwide system for assessing quality of education which should become an inseparable part of continuos education infrastructure allowing for the unity of educational space and increase in objectivity of assessment procedures for an education level through their separation from the processes of training and preparation. This system may be supplemented by non-commercial organizations holding exams and certification of the citizens' education potential. These structures should organize exams after secondary and high school while getting to the second level of higher education.

Modern research enabled us to make a conclusion that in the process of continues professional education it is vital to reconstruct the system of primary and secondary professional education institutions. Formally various types of primary and secondary professional education institutions commonly realize a similar set of educational programs. Thus, the structure of professional education is being destroyed. Meanwhile, the loss of close ties with enterprises and organizations and aging of material and technical resources in the institutions does not let them guarantee the quality of graduates' preparation necessary for modern economy. Moreover, this system is burdened by social commitments. Institutions of this level concentrate a greater part of troubled youth, so the solution of social tasks often dominates in the process of highly-qualified staff training.

Most employers highlight the great deficiency in working staff accompanied by business expenditures caused by low-quality education.

With a view of providing modern quality staff preparation in the system of professional education we propose the following:

1. Creating conditions for productive interaction of enterprises and educational institutions to organize and manage study process based on the state-of-the-art technological base.
2. Distinguishing between the mechanisms of students' social support, providing general education and organizing professional training by working out and implementing mechanisms of separate financing

of comprehensive syllabi and professional syllabi in institutions of professional education.

3. Creating organization and legal conditions for integrating educational programs of primary and secondary professional education into the system of continuous professional education together with spreading the module structure of programs for professions and specialties.

The law "On Education" describes the system of education as total of interacting successive educational programs and state education standards, as network of educational institutions realizing these programs and administrative bodies. This concept highlights not the organization and structural basis, as before in the centralized system of education, but, first of all, its content basis. Such understanding dictates the purposefulness of content and structural approach to constructing the system of continues education, supposing the priority of constructing the content of continues education over its organization forms. The concept of continues education may be referred to three objects (subjects):

- **Personality:** In this case it means that a person studies constantly either in educational institutions or in a self-instruction mode. This allows for three vectors of a person's movement in educational space. First, a person, staying at the same formal educational level, e.g. as a mechanic, a doctor or an engineer can improve his/ her qualification and professional skills. It may be conventionally called "forward movement vector". Secondly, ascending the steps and levels of professional education realizes the "vector of upward movement". With that, a person may gradually mount the levels of education or skip some of them. For example, a future student can continuously get primary, secondary and higher education or go to get higher education immediately after school. Thirdly, continues education supposes the opportunity not only of proceeding with, but also the change of the education major, presenting a possibility for an education maneuver at various levels of the lifetime according to the personality's demands, opportunities, social and economic conditions in the society. This is the "vector of horizontal movement".
- **Educational Processes (Educational Programs):** Continuity in the process of education is characterized by succession of the educational activity content in the process of transition from one kind of training to another, from one life stage to another;

- **Educational Institutions:** In this case continuity characterizes the network of educational institutions and their inter-connection creating the space of educational services able to satisfy all the variety of educational demands in the society as a whole and for every person in a certain area.

Presented below are the principles of continues education development which were separated on the basis of the dialectics category "content – form".

The content aspect, in its turn, is subdivided into two components: the "content" subsystem composition and its structural ties.

We find it important to highlight the principles of building content of continuos education corresponding to various vectors of a person's movement in the educational space:

- The principle of multi-level professional educational programs which supposes the presence of numerous levels and situations in professional education (the "upward movement vector");
- The principle of compensability of basic and postgraduate professional education (the "forward movement vector");
- The principle of study programs flexibility (the "horizontal movement vector").

Another direction of realization is considering the continuity of professional education as a system of educational processes (study programs) targeted at providing further development of the specialists according to their personal needs and social and economic demands of society.

We accept the following principles:

- The Principle of succession of educational programs which supposes their compatibility and correspondence to each other backed up by standartization of educational programs;
- The Principle of integration of professional education programs integration. Many lyceums and colleges realize study programs of primary and secondary professional education, often on 20-25 various specialties.

Thus, the integration of professional education subsystems and their organization structures seems inevitable contributing to the transformation of professional educational institutions into multi-level, multi-step and multi-multi-disciplinary institutions.

Otherwise, the possibility of the contrary process cannot be excluded. One and the same program can be realized in educational institutions of various types. This principle of organization forms flexibility ensures maximum variability of training forms.

The research has shown that the current trends in professional education include multi-disciplinarily and the structure with multiple levels. The education system comprises the following elements:

- "Multi-disciplinary" comprehensive schools are "common" schools, schools with in-depth study of certain subjects, lyceums, gymnasiums, etc. Apart from that, general education is realized according to majors in the institutions of primary and secondary professional education. A great degree of "specialization" is added to the content of secondary education by national, regional and local components;
- At least three steps in the institutions of primary professional education: training on the level of the first and second qualification labor grades; the level of primary professional education, qualification labor grades 3-4;
- The so-called "advanced level of primary professional education", qualification labor grades 4-5;
- At least two levels in institutions of secondary professional education: traditional and advanced, e.g. "technician" and "junior engineer" in technical colleges;
- In the institutions of higher professional education the system of higher education with 5 years of tuition and single qualification which was in action in previous years has been supplemented with a 4-year bachelor program and a 6-year master program;
- A 2-year educational program referring to the level of incomplete higher education has been introduced.

Compatibility of educational programs from the general education to postgraduate education is necessary for a student or a specialist to freely and confidently move in educational space through levels and steps (vertically) and stages and forms (horizontally).

Succession and compatibility mean that "coming out" from one stage of education must naturally "dock" with the "entrance" into the next stage. This necessitates transparent standardization of levels and steps of education is based on common goals of the whole system of continues education.

Multi-disciplinarily and multiple levels of educational programs are formed in institutions of general education, professional lyceums, colleges and universities. The paradox is connected with the fact that the links of education system, due to their traditional disconnection form multiple levels and multi-disciplinarily only for themselves, separately in general education, in professional education, in secondary and higher education. The institutions of postgraduate education, if any, work on their. As a result, with all the positive moments in growth of variety of educational systems and programs we face the situation of education space breakaway. Today, frequent are the cases when a student cannot be transferred from one school to another because different subjects are studied in different schools at different time even with the common state basic syllabus.

Many universities have started introducing entrance exams in the subjects not listen in the school syllabus. Additional entrance exams held by prestigious universities disorient general education schools and lead to the deformation of requirement for general education. Moreover, the results of the first university exams testify to the unsatisfactory selection of students.

The lists of specialties of training in primary, secondary and higher education do not agree with each other even in the names of professions.

So, lack of coordination in general educational programs on the all-state scale creates "educational dead ends", leads to other problems, particularly, conditions for corruption in education.

Nevertheless, succession of educational programs has begun to be formed partially, "from below" by the educational institutions themselves. Some schools, gymnasiums, professional lyceums, colleges make direct agreements with universities, create successive syllabi and programs and organize joint tuition of most capable students. There are numerous examples to this but the problem must be solved as a whole.

The above factors enabled us to make the following conclusions:

- In cultural and education environment of the state there have been formed the principles of continuous education system development;
- Partially built are multi-disciplinarily and multiple levels of educational programs, various forms of tuition are developing;
- Created is a network of educational institutions ensuring the succession of educational services.

Constructing a structural and functional model of the system of continues professional education for persons with health limitations means, firstly, creating conditions for its realization. The research on development of continues professional education and the structure of systemic theory of professional education development has revealed the main ideas of professional education evolution:

- Humanization of professional education (orientation to personality);
- Democratization of professional education (orientation to society and state);
- Advancing professional education (orientation to manufacture) is continuous professional education (orientation to system).

The data obtained from the research has allowed determining the main conditions for realization of the model of continues educations for persons with health limitations: humanization, democratization, advancement, continuity. These are the conditions necessary for the development of system of continuous professional education as a whole.

The problems of professional education humanization can be considered through humanization of education and fundamentally base, thus, necessitating the highlights on the general education component, the choice of either module or integration principle of content building, the transition to training the multi-discipline specialists, specification of basic training, strengthening the research potential and methodological preparation, socially-oriented and personalized and activity-oriented character of education and the teaching technologies used.

Consequently, the first condition for the system of the general model (set of models) of continues professional education is humanization which can be characterized as:

- Personal orientation of professional education for the persons with health limitations, the process and the result of the personality development, the means of its social and economic sustainability and social security in society;
- Modularity of professional education content;
- Strengthening the role of the independent education activity of the students with health limitations; providing freedom of choice of professional educational trajectories for students with health limitations;

- Formation of theoretical knowledge of the students with health limitations, with their practical professional demands and value orientations;
- Students' mastering the general competences for all kinds of professional activity (the skill of self-organization of study and professional activity, search for information, mastering new technologies, computer skills, foreign language knowledge, using databases, the knowledge of ecology, economics and business, financial knowledge, etc.).

The second condition for continues professional education is democratization considered as creating opportunities for each student with health limitations to realize own abilities in accordance with the demands of society and manufacture. Democratization of professional education supposes the necessity and possibility, and also means and conditions for realization of equal opportunities in getting professional education, its openness and variety of professional education institutions, cooperation of students and instructors, student self-administration, regionalization of professional education, international integration and cooperation, non-governmental forms of getting professional education with increasing authority at all levels.

Democratization supposes providing availability of professional education for all categories of youth and adults according to the principle of equal opportunities. Availability of education means a citizen's opportunity to get the quality education he/ she needs. Availability also means equality of education opportunities maintained through the following:

- Development of open distance education, extenuate, etc.
- Coordination of educational programs (the opportunities for mastering the new content, i.e. adequacy of educational standards);
- Social availability of professional education which includes both socially determined demands and traditions of education, and society's attitude to providing educational opportunities for persons with health limitations;
- Social partnership by means of attracting social professional associations to creating educational standards, organization of professional educational programs, to the assessment of professional education quality.

The advancing character of professional education can be considered as a condition for the future sustainable development of the country, its economy and social sphere and means reconsidering the content of professional education and technologies from the viewpoint of personality self-development, increase of population's education level, labor market monitoring, staff training in perspective directions, taking into account geographical location and perspective of the region's and the country's development as a whole.

Consequently, the third condition for continues professional education is the advancing continues professional education which means the following: the level of development of the system of general and professional education should be advanced and go ahead and form the level of personality development considering the advanced social order, implying forecasting, formation and advanced satisfaction of society's future needs for staff training. Advanced continues professional education is directed to the personality's self-development – self-development of intellect, will, emotions and sense motor sphere of students with health limitations.

The fourth condition is the continuity of professional education process which is understood as a possibility of the personality's multi-dimensional movement in educational space which means transition from the construction "education for lifetime" to the construction of "education through lifetime". Special interest presents vertical continuity (growth up the steps and levels of education with the change of education status). Continuity of professional education is ensured with the continuity of education content, education process, education organization, and also with flexibility of educational programs. The last condition can be realized with the module construction of proffessional education programs for their content to comply with other programs, levels and steps; creating conditions for parallel training in various educational institutions, in different programs of different levels.

For the student to freely move in educational space of continues education system, it is important to ensure agreement and compatibility of educational programs, their multi-level and successive character. Continuity in educational process becomes a characteristic of the personality's inclusion into education process at all the stages of its development. It is also linked to the succession in educational activity in transition from one its kind to another, from one life stage to another.

The fifth condition is diversification of professional education which supposes creating conditions for diversity of educational trajectories backed up by an unlimited number of educational programs considering individual characteristics and personal abilities within the framework of the new typology of professional educational institutions.

The sixth condition is the information of professional education which means development of modern information technologies and high-speed communication lines which expands the opportunities of using these technologies in professional education and research. This makes it necessary to develop territorial networks of data communication meant for maintaining educational organizations' access to Russian and international information networks, perfection of net infrastructure and information content, creating conditions for students' access to world information resources. This should result in the common information educational network in the region allowing for the use of cutting edge software for exchanging various data – from texts to video conferences, applying interactive multimedia technologies in education process. This will also make it possible to form inter-regional data of labor market, analyze perspective demands for professions and specialties.

The seventh condition is the integration of science and education:

- Working out mechanisms for formation and stimulation of solvent demand for high technologies and results of scientific research, selection and realization of most efficient achievements of science and technology on the basis of the federal contract system;
- Concentration of resources on priority directions of science and technology development; structural reconstruction of scientific research front, their priority support by the state in the spheres connected with the necessary and long-term vital interests of a person, society and state;
- Creating infrastructure providing for conducting and use of results of such research and their transformation into socially beneficial scientific and technical solutions, commercially efficient, ecologically safe and competitive technologies, commodities and services;
- Introduction of the system of effective stimuli on federal and regional levels for increasing the prestige of research, personnel retention, attraction of the young specialists to scientific, technical and innovative spheres; formation of necessary organization, economic, legal, information and other kinds of research activity of various centers and institutions with the view of approaching sustainable development of society.

The integration of science and education allows for development of scientific background of education system, to strengthen the potential of university research, to solve the tasks of staff training for innovative activity.

Working out the model of continues professional education necessitates analyzing problems in the sphere of continues education. The analysis of official statistics and the data of the questionnaires for enterprise management in the real economy sector made it possible to single out the trouble issues in education sphere negatively affecting further development of economy and business:

- Reduction of labor resource's growth resulting from the negative demographic situation in the period till 2015 (twice approximately);
- Difficulties in supplying real manufacture with qualified workers and specialists due to their deficiency in the labor market;
- Noncompliance of professional training of workers and specialists with the labor market demands;
- Absence of common strategy in staff training and realization of continuos education;
- Absence of the common system of certification for students and specialists.

Professionally competent specialist is characterized by a number of qualities:

- Education (knowledge, skills, intellectual interests, strives for and ability to constantly perfect knowledge, worldviews); social training (moral, aesthetic, physical, labor);
- Socialization (readiness for active professional and social activity, self-realization); culture (exteriorization of cultural values of humanity, culture of labor and communication);
- Articulated individual features (creativity, analytical skills, formed memory and thinking, etc.).

Consequently, at each stage of professional development one should consider not only formation of certain skills and knowledge, but also building up systemic qualities of the professional's personality.

Creating the common model consisting of separate interconnected models for the system of continues education is connected with solution of various tasks, among which are the following ones:

- Training highly qualified workers and specialists for most-demanded specialties on labor market in accordance with the region's interests;

- Providing for faster and more flexible adjustment of professional education system to labor market demands and changes;
- Compatibility of professional training levels within the framework of common educational programs;
- Multiple levels and openness of educational space;
- Completeness of each level of professional education (getting specialty, acquiring qualification); transition from one level of tuition to another according to the results of knowledge control on competitive basis;
- Diversity of forms and methods of training, creation and exploitation of new pedagogical technologies based on modern information and telecommunication resources;
- Providing for variability of educational process;
- Orientation to the development of fundamental and applied research as inseparable part of university specialist' training;
- Establishing close relationship with enterprise and labor market;
- Succession in the study of disciplines in professional education programs of different levels;
- Intensification of the process of the students' professional self-identification.

Construction of the model of the system of continues professional education makes it necessary to consider the previous experience in the system of education for persons with health limitations.

Educational practice reveals that most models of continuous professional education are regionally-oriented and realize career-orientation, training and profession-adaptation functions. The main idea of constructing such models is linked to the organization of education process, supposing the interaction of all the steps of professional education with each other and sectorial enterprises ensuring stimulation of students' professional self-determination, letting them build up an individual trajectory at any stage of professional and even pre-university training. In this case, professional orientation wills guarantee striving for efficiency of training and successful adaptation of young specialists in enterprise.

Methodological grounds of constructing models for continues professional education also present certain interest.

Most authors determine methodological approach as synergetic, resource, personality-oriented, context and competence approaches.

From the viewpoint of synergetic approach the modules of the continues professional education system are characterized by:

- Openness (determined by the links between educational institutions, enterprises and their environment);
- Dynamics (the model continuously adapts to the changing conditions);
- Ability to resist the outer destabilizing pressure.

Working out the model complies with the logics of resource approach which means creating necessary conditions for its realization. In reference to the student's personality as a subject of educational process in conditions of continues professional education there exist outer and individual resources. Considering the students' individual resources is based on their revealing, efficient use and providing conditions for their development. Outer resources include legal basis, material, technical and personnel-related background of the model.

The legal base for continues professional education model comprises:

- Legal documents of the ministry of education, regulating realization of the concept of continues professional education;
- Regional program of professional education development;
- Agreement on state corporate partnership in the sphere of professional education, professional training, re-training, professional development in the system of professional education;
- Agreement on cooperation of educational institutions with enterprises;
- Agreements of the university and schools on creation of profile classes;
- Agreements of the university with educational institutions in the sphere of professional training.

The functioning of the continues professional education model supposes involving into study process the teaching staff of general education institutions, university faculty, highly-qualified specialists from the allied spheres of industry and economy, heads of state and municipal bodies, guest professors from other regions and foreign faculty.

Material and technical support of the model in the study process supposes using laboratories of educational institutions, instructional classes, industrial platforms, etc. Thus, the model may be defined as an integral system developing through active use both of internal resources, and external opportunities of the environment. The essence of the model organization according to resource approach is concentrated around creating the conditions for its subjects' interaction which will ensure the development for each one.

- In creation of the general model we have used the personally-oriented approach according to which priority is given to the value and emotional sphere of personality, its activity and personal position. Main ideas of the personally-oriented approach as applied to construction the general model are as follows (Mkrttchian & Belyanina, 2016) The study process is filled with new personal ideas, values, relationships and supports the individuality and uniqueness of each student;
- Tuition is directed to formation of the vitally important knowledge, necessary for continues development;
- The learning process reveals the subject experience of each student;
- Tuition technologies consider the students' education abilities and the specificity of their major.

Another approach, the context one, is aimed at the students' preparation in accordance with their future profession. It supposes the student's personal inclusion into the processes of cognition and mastering the future profession and directly determines his/ her successful activity in the future.

The realization of the competence approach is linked to the urge to define the vital changes in professional training as determined by the transformation in society and economy. New quality of education is closely connected with the grounded forecasting, projecting, modeling of necessary competences of the student (Mkrttchian & Belyanina, 2016).

1. The basic principles of the model under consideration are as follows (Mkrttchian & Belyanina, 2016):Rationality supposing orientation in education organization and content to the local conditions and labor markets, dependency on concrete economic and social conditions of region development; the graduate's expanded professional profile as a feature in high demand in the market;
2. Integration supposing inclusion of all the system's elements into the process of professional training with the view of rational use of resources;
3. Succession consisting in the corresponding of the graduates' qualification characteristics to the demands of the customer enterprises;
4. Intensification of education consisting in raising efficiency of tuition by means of new technologies, forms and methods in educational process;
5. Differentiation expressed in realization of the students' right for choosing an individual educational and career trajectory;
6. Variability of educational programs providing the student with the free choice of individual education trajectory.

It is common knowledge that the staff professional training is based on the use of various technologies (personality-oriented, cooperation technologies, project and activity training); educational methods (problem method; searching method; the method of solving definite situational tasks; the method of doing creative tasks with research elements; project method); forms of study (individual; group; frontal method; self-education).

The research on integration in projecting educational systems showed that integration due to its dialectical and logical unity allows for reconstructing the unity of processes and systems necessary for solving education tasks. Integration approach is of great importance while building the model of specialists' training. It is determined by the development of the modern national economy as part of the world economy (Mkrttchian & Belyanina, 2016) . Changing conditions of professional activity served the reason for a number of contradictions reflected on the sphere of education, particularly, for the discrepancy between changing demands for specialists in developing economy and reflecting new requirements in educational standards; between the market's need for specialists with practical professional background and the absence of this experience in graduates. These contradictions can be solved if the specialists' training is based on integration of professional education and professional practice. With that, the integration can be described as different stages of a single process: preparation for activity and activity itself.

The common ground for an individual's professional development is formed by means of social integration of the structure "school-university-postgraduate school" at the points "school-university" (lower step of professional training) and "university – postgraduate school" (higher step of qualification) – practical and research activities.

Social integration of school and university on the lower step of professional training provides for the continuity of going from one stage of personality development to another. Quantity accumulation of general knowledge, development of personality qualities and an individual's gnostic abilities leads to their quality transformation in professionally-oriented space. Strategic partnership contributes to realization of own functions for all the participants of study process which lets the partners to solve their problems. Thus, school can created a database for the students' self-determination in their further development based on the information of professional market, the role and place of professionals in national economy development.

Psychologically, professional orientation contributes to an individual's readiness for further learning by means of raising motivation for mastering professional skills.

The changing character of education, increase in the amount of self-study, activation of training and organization of stage communications allow for gradual introduction of the new tendencies in secondary school.

At the point of the university's transition to the practical activity sphere (here also refers the stage of post-graduate study) the integrated space is characterized by a brand new stage in personality development. It is determined by the discrepancy in the leading activity of the individual: between accumulation of professional knowledge and demand for their realization. This discrepancy is solved by means of transformation from the study to professional practice which proceeds with the process of the specialist's professional development.

The dialectics of development supposes the presence of some variants of entering the practical activity sphere: development follows a number of ways, not one. The dialectic ontogeny process continues at the postgraduate stage while combining research, teaching and practical activities.

The specialists from the sphere of professional activity return to this stage as development is not a straight line and not a move around the circle but a helix with a number of whorls. The cognition process sometimes repeats the past cycles but always on the new basis. The step of research activity in the system "school – university – postgraduate school" is determined logically and dialectically and presents the integration of study, research and practical activates.

Social integration of the university and organizations from the practical spheres lets the university flexibly structure the content of education with orientation to new scientific outcomes and demands of professional practice and labor markes. Internships bring corporate culture of strategic partners in the study process. In the process of tuition the students have the possibility of dealing with the strategic partners' corporate culture, of studying the rules of behavior in professional sphere and determining the potential workplace before graduation.

Connection with the corporate structure allows for development through educating staff, their getting additional qualifications in flexible conditions and forms of tuition.

Therefore, the model of special and structural integration of professional education and professional economic activity reflects the methodology, theoretical and methodological prerequisites for preparation of specialists with economics major. The key element of the model is the integration tools: internship bases, joint enterprises, workshops, students' professional activity. The mechanism of social integration of education and professional activity spheres is strategic partnership creating the field for developing the specilist's personality development in the common space of living. The social integration mechanism ensures the correspondence of the educational structure "school – university – postgraduate school" with the external environment (professional internship) and, being an open system, is able to develop and renew oneself under the influence of changing conditions.

Taking into consideration the accumulated experience in professional education system lets us turn to constructing own model of continuos education system for the persons with health limitations that will have its specificity determined by the specifics of this category of learners.

The model of continues professional education for the persons with health limitations, in our view, should be worked out in accordance with approaches to and demands for continues education system highlighted in the normative documents regulating the functioning of continues education for persons with health limitations; should consider conditions of its realization and contain components defining the contents of continues education for persons with health limitations.

Constructing the model of the system of continues education for persons with health limitations requires determining its structural components: the system of requirements for education; approaches and principles of creating the system of continues education; conditions for model realization; content of continuous professional education; complex support; creation of the environment free from barriers; the system of extracurricular activity; links of professional education institutions with labor market. Below are the characteristics of all the structural components of the model of continues professional education system for persons with health limitations (Mkrttchian & Belyanina, 2016). The first structural component of the model is the system of requirements. Normative documents prioritize the values of world and Russian culture, humanistic morals, civility, worldviews and methodological solutions, oriented to formation of new generations of specialists capable of creative, professional, responsible activity; provision of equal educational opportunities for persons with health limitations (Mkrttchian & Belyanina, 2016) .

The above enables us to determine requirements for the general model of the system of continues education for persons with health limitations: providing equal opportunities and conditions for persons with health limitations to get professional education aimed at training competitive specialists capable of responsible activity, oriented to general cultural and moral values, humanistic morals and civility.

- The second structural component of the general model of the system of continues a professional education for persons with health limitations is determining approaches to its organization. Generally, the main approaches are defined as an approach to the system of continues education for persons with health limitations (Mkrttchian & Belyanina, 2016) The aim of social development is a person's continues moral, personal and professional development;
- Continues education should provide each person with the institutional opportunity of forming an educational trajectory and of getting the professional training which is required for his/ her further professional, career and personal growth;
- Creating infrastructure of access to continues professional education throughout the period of professional activity.

The main approaches to constructing the general model of the continues professional education system may be summed up as follows: creating availability of professional education for persons with health limitations through formation of individual educational trajectory contributing to continues moral, personal and professional development in accordance with their abilities and possibilities.

- The third structural component of the model of the system of continues professional education for persons with health limitations is defining the principles of professional education organization. In a general way, they may be presented as (Mkrttchian & Belyanina, 2016): Basic property, multiple levels, diversification, complementarity, flexibility, successive integration, availability;
- The possibility of changing the major at different life stages taking into account personality's demands and needs as well as social and economic conditions in society;
- Succession of educational activity content in transition from one kind to another, from one life stage to another;
- The space of educational services capable of satisfying all the number of educational needs arising in society as a whole, in a separate region as well as individually.

The above enables us to formulate the basic principles of constructing the model of continuous professional education for persons with health limitations: satisfying educational needs, in accordance with their possibilities, as based on multiple levels, diversification, complementarity, flexibility and succession of integration, flexibility and availability of the system of continues professional education.

The fourth structural component of the general system of continues professional education for persons with health limitations are conditions for realizations of continues professional education model (Mkrttchian & Belyanina, 2016).

Hereinafter mentioned are the main conditions for realization of the general model of continuous professional education:

- Personal orientation, differentiation and individualization of education backed up by state educational standards on the basis of diversity and variability of educational institutions;
- Creation of information environment in each educational institution and wide use in educational process of advanced pedagogical technologies; development and acceptance of corresponding norms of their realization;
- Integration of study programs of primary, secondary and higher professional education; spread of the module structure of programs of preparation in professionals and specialties;
- Creation of new mechanisms of assessment and accreditation of educational programs involving social and professional organizations, employers 'associations.

Conditions for realization of the model of continuous professional education system for persons with health limitations: personal orientation, differentiation and individualization of education as based on the use of advanced educational technologies, working out of the module structure of integrated educational programs, information of educational space and involving community into the process of the graduates' quality assessment.

The fifth structural component of the general model of the system of continuos professional education for persons with health limitations is the content of education (Mkrttchian & Belyanina, 2016).

The content of continues professional education includes:

- Continues professional training of the staff which is based on the use of various teaching technologies, teching methods (problem method, search method, method of solving definite practical tasks, method of doing creative works with research elements, project method), forms of tuition (individual, group, etc.);

- Development of open distance education, externate, etc.
- Agreement of educational programs (the possibility of covering new content – adequacy of educational standards);
- Personal orientation of professional education of persons with health limitations as means of its social sustainability and economic protection;
- Formation of theoretical knowledge of persons with health limitations in accordance with their practical professional needs and value orientations;
- Free choice of professional individual educational trajectories for students with health limitations;
- Modularity of professional education content;
- Students' mastering the competences common for all kinds of professional activity (skills of self-organization of study and professional activity, search for information, mastering new technologies of activity, computer skills, foreign language knowledge, data using skills, the knowledge of ecology, economics and business, etc.);
- Social availability of professional education which includes both socially determined needs and traditions of getting education, and the attitude of society and state to providing educational opportunities for persons with health limitations;
- Changing character of study, increase in the volume of independent work, activation of study activity and organization of step communications.

The content of the model of continues professional education system may be summarized as a process of formation of students' key competences on the ground of choosing individual educational trajectories, forms of study according to their possibilities, needs, values and personal orientation with a view of their social and economic sustainability and social security in society.

The sixth structural component of the model of continues professional education system for persons with health limitations is the complex of social and medical-psychological-pedagogical support including (Mkrttchian & Belyanina, 2016):

- Creation of integrated educational space as a complex of conditions for developing the personality's potential on the basis of implementing modern educational technologies including health-protecting, information technologies, pre-profile preparation and profile tuition;

- Maintenance of physical, psychic and social health, timely diagnostics and correction, systematic medical and psychological, pedagogical and social support for the students with health limitations;
- Individual program of rehabilitation;
- Career-oriented work for choosing the future specialty for persons with health limitations;
- Psychological and medical support for students with health limitations (psychologists, medical workers, social teacher).

Characteristics of complex social and medical psychological and pedagogical support for the persons with health limitations can be determined as a system of professional activity of different specialists on creating conditions for the subject's accepting optimal solutions for personality development through medical and psychological support, pedagogical support and professional adaptation.

The seventh structural component of the system of continues professional education for persons with health limitations is barrier-free environment (Mkrttchian & Belyanina, 2016):

- Required planning of study rooms, rampant and elevators;
- Equipping with informational technologies and adapted facilities (vision and hearing pathologies), free Internet access;
- Architecturally-available environment and special equipment (for students with loco motor system problems); equipped medical and rehabilitation rooms; equipped places for recreation;
- Study process is compliance with sanitation standards.

Characteristics of barrier-free environment for persons with health limitations are concentrated on creating conditions for equal opportunities and availability of professional education as backed up by modernization of educational institution infrastructure.

The eighth structural component of the model of the system of continues professional education for persons with health limitations is extracurricular pedagogical activity (Mkrttchian & Belyanina, 2016). Theoretical background allows for stating that the main content of the university pedagogical activity comprises inclusion of a student with disability into educational and social life at university; application of personal growth programs with the use of active training methods; inclusion of students into student interest and hobby clubs, sport activity.

The main content of extracurricular activity of persons with disabilities is concentrated on inclusion of students with heath limitations into socially important cultural, creative and sport activity aimed at building their social and cultural competences for integration and adaptation in society.

The ninth structural component of the model of the system of continues professional education is connection with labor market expressed through (Mkrttchian & Belyanina, 2016):

- Mechanism of social integration of spheres of education and professional activity realized in strategic partnership creating the ground for developing the specialist's personality in the common space of its life activity;
- Integration of university and practical sphere organizations allowing for the university structuring the content of training in accordance with new research and professional practice as well as labor market demands.

The content of interacting with labor market is concentrated on forecasting, projecting, modeling of key competences for students with health limitations on the ground of changes taking place in society and demands for quality of education and professional training of persons with disabilities. Figure 1 shows the scheme of the general model of continues professional education for persons with health limitations consisting of worked out models (Mkrttchian & Belyanina, 2016).

Where block 1-1 shows the structural component of the general model generalizing "Demands of society, employers and students for professional education for persons with disabilities ", block 1-1 shows characteristics of structural component in general model, which generalizes "Ensuring equal opportunities in getting professional education for persons with disabilities aimed at training competitive specialists capable of professional responsibility, oriented to cultural and moral values, humanistic morals and civility", block 1-1 shows criteria of the general model "Share of students with disabilities having access to the system of continuos professional education".

Block 2-1 shows the structural component of the general model summarizing "Approaches to getting continues professional education for persons wih disabilities". Block 2-1 illustrates the characteristics of the structural component in general model which generalizes "Creating availability of professional education for persons with disabilities through formation of an individual educational trajectory providing for continues moral, personal

Figure 1. The diagram a of the general model of continuing professional education of persons with disabilities Act (IDEA), consisting of designed models

Stages	Goals	Tasks
Preparatory stage	Block 1-1. Creating the normative basis for ODL use. Creating technical basis. Block 2-1. Creating course packet (of the net educational resource). Training the staff	- Formulating the goals of ODL usage, working out the model of DL in accordance with normative acts. - Defining the terms of educational process launch. -Creating the technical base for DL - Selecting/ developing the net educational resource. - Defining the principles of forming the study groups. - Defining the terms and succession of the courses. - Training net instructors, mentors, organizers, etc.
Stage 1 Realized before the academic year	Block 3-1. Questionnaire survey, revealing the students' educational needs. Block 4-1. Conducting the normative basis for ODL. Block 5-1. Compiling individual study plan. Compiling and correcting the students' study schedule. Block 6-1. Compiling the thematic outline.	- Study and analysis of the resource and course content - Studying the students' individuals needs. - Compiling individual study schedule considering the major and the study plan. - Meeting the study group -Compiling thematic schedule including: • preliminary defining the goals for each topic; • defining the academic hours for each topic; • defining the place of project and research activity; • a more detailed analysis of the program content; • defining the lesson types on the given topic including the venue for laboratory works, practical lessons, etc. • defining the venue for online lessons and their types; • defining the venue for discussions and conferences (both online и offline)
Stage 2 Realized before the start of each lesson and every topic	Block 7-1. Teacher's preparation for a certain lesson	• defining the goals of studying each topic in accordance with the major - defining the content of additional theoretical material on the given topic and preparing the list of sources of supplementary reading for students - Analysis of the system of tasks offered for solution in resource; defining the degree of its completeness, complexity of the offered tasks, the system of differentiating the tasks; its correspondence to the formulated goals. - Defining the necessity in additional set of tasks
	Block 8-1. Drafting instruction for the lesson Block 9-1. Working with home assignmentsP	- Planning the system of control and self-control of the students' mastering the content of the topic and achievement of the goals set - Selection of additional tasks to be solved in class and at home - Checking home assignments, assessment and commenting on home assignment

and professional development of persons with health limitations, taking into account their possibilities and abilities", block 2-1 illustrates the criteria of the general model "Share of students with disabilities learning individually, in a distance mode".

Block 3-1 shows the structural component of the general model summarizing "Principles of continues professional education for persons with disabilities", block 3-1 illustrates the characteristics of the structural component in general model which generalizes "Satisfying educational needs of persons with disabilities in accordance with their possibilities on the ground of multiple levels, diversification, completion, flexibility, succession, integration and flexibility and availability of continues professional education system", block 3-1 shows the criteria of the general model "Share of students with disabilities getting additional education. The level of satisfied educational is needs".

Block 4-1 illustrates the structural component of the general model summarizing "Conditions for realization of continues professional education for persons with disabilities", block 4-1 shows the characteristics of the structural component of the general which is generalized "Personal orientation, individualization of education for persons with disabilities based on using advanced technologies, working out the module structure of integrated educational programs, information of educational space and involving community into assessing the quality of graduates' preparation", block 4-1 shows such criteria of the general model "Developed methodological support for individually-oriented study processes of persons with disabilities".

Block 5-1 illustrates the structural component of the general model generalizing "Content of education of the system of continues professional education", bloc 5-1 shows the characteristics of the structural component of the general model which is summarized in "The process of building up the students' key competences on the basis of choosing the study content, individual educational trajectories and forms of tuition in accordance with their possibilities, needs, values, personal orientation with a view of providing their social and economic sustainability and social security in society", block 5-1 shows the criteria of the general model "Share of the students with disabilities included in various forms of education. High index of quality of education for persons with disabilities is during intermediate and final assessment".

Block 6-1 shows the structural component of the general model summarizing "Complex social, medical, psychological and pedagogical support of persons with disabilities", block 6-1 shows the characteristics of the structural component of the general model summarizing "The system of professional activity of various specialists creating conditions for the subject's taking optimal decisions for the development of the personality with disabilities through providing medical, rehabilitations, psychological support, pedagogical support", block 6-1 shows the criteria and indices of the general model "Share of the students with disabilities involved in complex social, medical, psychological and pedagogical support".

Block 7-1 shows the structural component of the general model summarizing "Barrier-free environment", block 7-1 shows the characteristics of the structural component of the general model summarizing "Creating conditions for equal opportunities and providing availability of professional education for persons with disabilities on the basis of modernized structure of the educational institution", block 7-1 shows the criteria and indices of the general model "Level of the students with disabilities' satisfaction with availability of educational environment and infrastructure".

Block 8-1 illustrates the structural component of the general model summarizing "Extracurricular activities", block 8-1 shows the characteristics of the structural component of general model summarizing "Inclusion of students with disabilities in social, cultural, sport activities aimed at building their sociocultural competences for integration and adaptation in society", block 8-1 shows such criteria and indices of the general model "Share of the students with disabilities involved in social, cultural, sport activities".

Block 9-1 shows the structural component of the general model summarizing "Links to labor market", block 9-1 shows the characteristics of the structural component of general model summarizing "Forecasting, projecting, modeling key competences of students with disabilities according to changes in society and requirements for the quality of education and professional training of persons with disabilities", block 9-1 shows such criteria and indices of the general model "The number of social partners and employers involved in assessing the quality of education of persons with disabilities". Figure 2 shows the structural and functional model of the system of continues education in sliding mode for persons with disabilities for solving tasks 1 through 7 with the use of the corresponding approaches and principles covered in Mkrttchian et al. (2016)

- **Task 1-1:** "Training of highly-qualified workers and specialists in most demanded specialties in labor market in accordance with the region's interests " is solved by means of approach and principle "Personal sense: realized tuition should be filled with personal senses, values and relationship of the person with disabilities".

Figure 2. Structural and functional scheme of the general model of continuous education of persons with disabilities controlled intelligent agents in the sliding mode

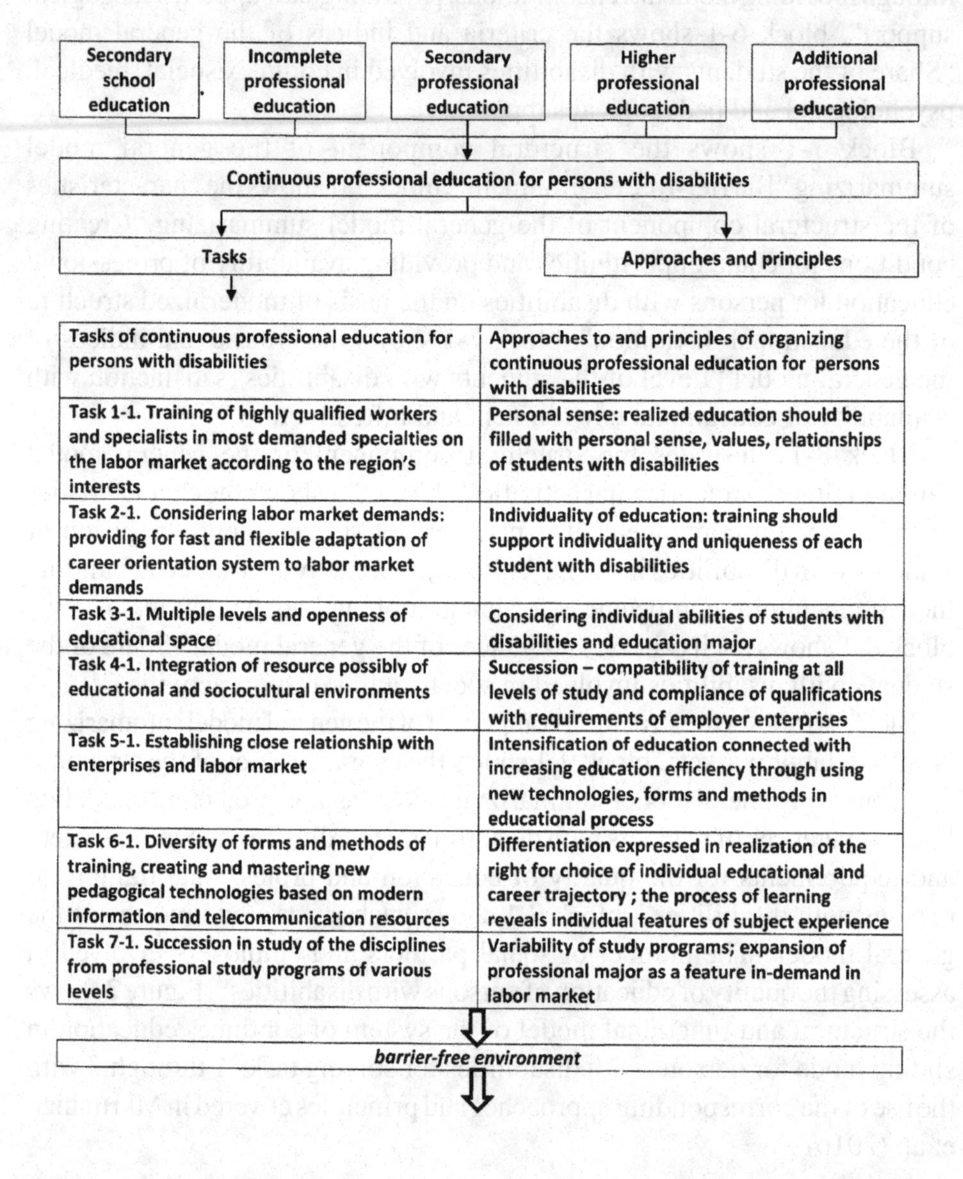

Tasks of continuous professional education for persons with disabilities	Approaches to and principles of organizing continuous professional education for persons with disabilities
Task 1-1. Training of highly qualified workers and specialists in most demanded specialties on the labor market according to the region's interests	Personal sense: realized education should be filled with personal sense, values, relationships of students with disabilities
Task 2-1. Considering labor market demands: providing for fast and flexible adaptation of career orientation system to labor market demands	Individuality of education: training should support individuality and uniqueness of each student with disabilities
Task 3-1. Multiple levels and openness of educational space	Considering individual abilities of students with disabilities and education major
Task 4-1. Integration of resource possibly of educational and sociocultural environments	Succession – compatibility of training at all levels of study and compliance of qualifications with requirements of employer enterprises
Task 5-1. Establishing close relationship with enterprises and labor market	Intensification of education connected with increasing education efficiency through using new technologies, forms and methods in educational process
Task 6-1. Diversity of forms and methods of training, creating and mastering new pedagogical technologies based on modern information and telecommunication resources	Differentiation expressed in realization of the right for choice of individual educational and career trajectory ; the process of learning reveals individual features of subject experience
Task 7-1. Succession in study of the disciplines from professional study programs of various levels	Variability of study programs; expansion of professional major as a feature in-demand in labor market

- **Task 2-1:** "Considering labor market demands: the system's flexible adaptation of the system of professional education to labor market demands" is solved by means of approach "Individuality in education: education should support individual and uniqueness of each student with disabilities".

- **Task 3-1:** "Multiple levels and openness of educational space" is solved by means of approach and principle "Considering individual opportunities of students with disabilities and specifics of the major".
- **Task 4-1:** "Integration of resource opportunities of educational and sociocultural space" is solved by means of approach and principle "Succession as compliance of training in all educational levels and the correspondence of the graduates' qualification characteristics to the requirements of the enterprises".
- **Task 5-1:** "Establishing close links with industries and labor market" is solved by means of principle "Intensification of education consistent in increasing efficiency of learning through the use of new technologies, forms and methods in educational process".
- **Task 6-1:** "Diversity of forms and methods of training, creation and application of new pedagogical technologies based on modern informational and telecommunication resources" is solved by means of approach and principle "Differentiation expressed in the students' realization of the right for choosing the individual educational and career trajectory; the study should reveal each learner's subject experience".
- **Task 7-1:** "Succession in studying disciplines in professional educational programs of different levels" is solved by means of principle "Variability of educational programs; expansion of professional major as a demanded feature of the labor market".

CONCLUSION

Sliding mode can serve the ground for developing the model of continuous education for persons with disabilities. The sliding mode technique allows for taking decisions in invariant conditions, analysing the situation, evaluating the risks and, as a result, taking decisions in control and organization in the course of tuition. The system of continuous education is getting adapted to the requirements of the normative documents and the current trends, functions according to the logic of the sliding mode.

REFERENCES

Mkrttchian, V., & Belyanina, L. (2016). The Pedagogical and Engineering Features of E- and Blended Learning of Aduits Using Triple H-Avatar in Russian Fedreration. In V. Mkrttchian, A. Bershadsky, A. Bozhday, M. Kataev, & S. Kataev (Eds.), *Handbook of research on estimation and control techniques in e-learning systems* (pp. 61–77). Hershey, PA: IGI Global. doi:10.4018/978-1-4666-9489-7.ch006

Mkrttchian, V., Bershadsky, A., Bozhday, A., Noskova, T., & Miminova, S. (2016). Development of a Global Policy of All-Pervading E-Learning, Based on Transparency, Strategy, and Model of Cyber Triple H-Avatar. In Developing Successful Strategies for Global Policies and Cyber Transparency in E-Learning (pp. 207-221). Hershey, PA: IGI Global.

KEY TERMS AND DEFINITIONS

Education Technology: Are technical, biological and engineering systems for Education whose components are combined, controlled and generated using the aligned single processing core. All the components at all levels of interaction are combined in the network infrastructure. All components include built-in calculators, providing data processing in real-time.

Indicator of Sliding Mode: The software for control virtual research space, maintain it sliding mode.

Moderator Avatar: Personalized graphic file or rendering that represents a computer user used to represent moderator in an online environment.

Researcher Avatar: Personalized graphic file or rendering that represents a computer user used to represent researcher in an online environment.

Triple H-AVATAR Technology: The technology of modeling and simulation based on known technology of Avatar used in the HHH University since 2010.

Virtual Research Environment: The space where with the help of virtual reality creates a special environment for research.

Chapter 5
Terms of Adaptive Organization of the Educational Process of Persons with Disabilities with the Use of Open and Distance Learning Technologies (Open and Distance Learning - ODL)

ABSTRACT

Modern world is characterized by the high rate of life, work and learning. Higher education is becoming a necessary condition for competition in labor market, successful business, and financial sustainability. To comply with new requirements, modern education should be available, be based on the use of electronic learning models with application of innovation methods and technologies including open and distant education. ODL is a generic term which defines many other practices such as distance education, distance learning, open learning, e-learning, m-learning, virtual learning, online learning, educational technology and learning technology. The globe has

DOI: 10.4018/978-1-5225-2292-8.ch005

been witnessing an age in which change is an important factor and ODL is not immune to these emerging changes. Therefore, ODL should embrace the fundamental changes to survive in a rapidly advancing world. In this regard, one of the best strategies to survive and compete is to understand the administration and leadership in ODL, and identify future planning accordingly.

INTRODUCTION

In the situation of rapid development of information technologies ODL is one of the key elements allowing for maximum individualization and optimization of the study process. ODL is a system of forms and methods of learning organization enabling the student to get education irrespective of his/her location and distance from instructors with the use of information and telecommunication technologies. ODL provides for formation of the skills of creative thinking, effective search, selection, structuring and analysis of information. A new form of pedagogical control (interactive mode) contributes to constant stimulation of the student to evaluate him/her and correct educational activity.

ODL system has a number of advantages before the traditional form:

- Maximum individualization of the learning process;
- Flexibility: the possibility to study in any location without changing the usual mode of life;
- Autonomy: studying at any time and any place excluding long presence in the study rooms;
- Optimization of learning expenses;
- Efficiency and objectivity of evaluating the students' progress;
- Flexibility of the structure of education organization;
- Possibility of learning process intensification;
- providing for psychological and pedagogical support of the learning process;
- Differentiated/ individual approach to students.

ODL can be used: in everyday class system; in profile grades; for optional (special/ supplementary) courses; project and research activity; for preparation for final exams and entrance exams to university.

MAIN FOCUS IN CHAPTER

Solutions and Recommendations

The following didactic models of lessons with the use ODL can be offered: the model of multi-level complex lesson; the model of revision lesson; the model of generalizing lesson; the lesson aimed at filling the gaps in knowledge; profile lesson with the in-depth study of the material (Mkrttchian, et al., 2015a).

Controlling models refers to lessons of preparation for final exams and self-control; multi-level home assignments; thematic tests.

Multifunctional models cover laboratory works with remote access; Internet (Online)-conference / discussion; Online-dialogue; search model.

Information technologies are intended to make ODL an educational process with maximum convenience and efficiency for the students. Listed below are the skills and competences acquired in the process of ODL which define the person's further success in various spheres of activity:

- Goal-setting, taking decisions;
- Conscious choice and responsibility;
- Independent planning and activity organization;
- Working in information space: accurate wording of search demands;
- Mastering the skills of scanning reading and reading for detail;
- Presentation of activity results;
- Reflection;
- Development of visual and logical memory;
- Self-education.

ODL models suppose creation of centralized informational education resource as electronic basic and profile courses (Mkrttchian. et al., 2015a)

Highlighted below are the most important processes of ODL system.

- The processes of development and support of resources is development of content; support of working capacity and development of educational space; technical support of the study process.
- Pedagogical processes and processes of managing education are education planning; managing study groups; learning; education management.

- Organization processes: admitting students; solving organizational issues of the students; managing document flow and formation of attestation documents. ODL organization includes various stages solving various tasks and providing for achievement of certain goals (Figure 1).

Normative base of the educational process is the Law of education which says:

Competence of the educational institution includes the use and development of educational process methodologies, including distance education technologies. The educational institution has the right to use distance educational technologies in all forms of getting education in the order set by the federal (central) state body of education control. (Mkrttchian et al.,2015b)

Distance learning courses may be used in the following ways:

- As study subjects of the resource center;
- As study courses;

Figure 1. Block diagram of the organization: The Open and distance lifelong learning (ODLL) people with the intelligent agents in sliding mode

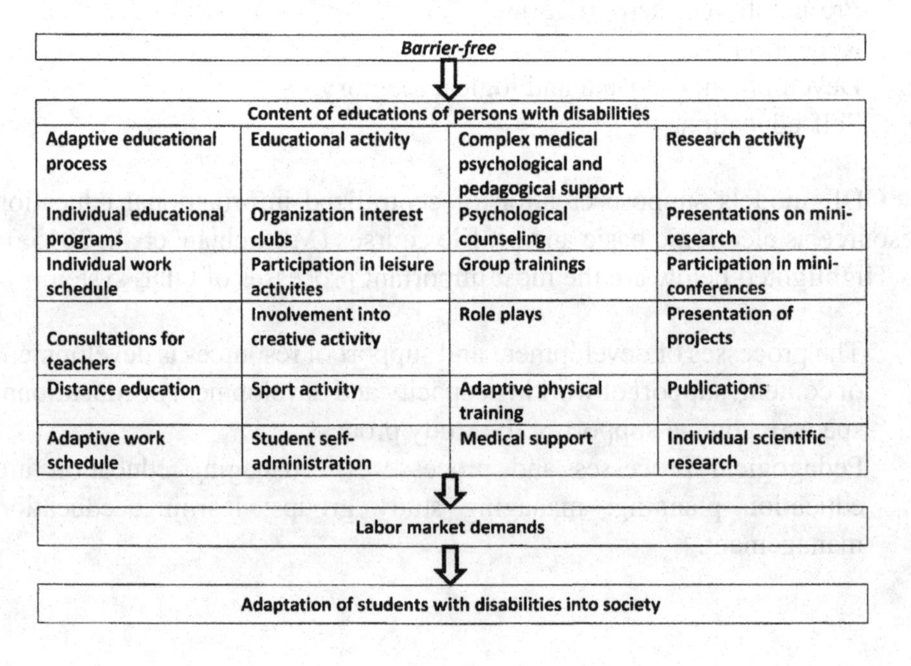

Barrier-free

Content of educations of persons with disabilities			
Adaptive educational process	Educational activity	Complex medical psychological and pedagogical support	Research activity
Individual educational programs	Organization interest clubs	Psychological counseling	Presentations on mini-research
Individual work schedule	Participation in leisure activities	Group trainings	Participation in mini-conferences
Consultations for teachers	Involvement into creative activity	Role plays	Presentation of projects
Distance education	Sport activity	Adaptive physical training	Publications
Adaptive work schedule	Student self-administration	Medical support	Individual scientific research

Labor market demands

Adaptation of students with disabilities into society

- As additional educational services for pay.

Creating Conditions for Persons With ODL

For the children with disabilities (who cannot attend educational institutions because of health limitations) there must be necessary conditions for learning according to the comprehensive or individual program at home. The order of training children with disabilities and compensation for their parents' expenses for these goals is defined by the laws or normative acts. The problems of ODL for children with disabilities in a certain territorial subject must be regulated by the laws and other legal acts of the state authorities. Organization of the Center for distance learning for children with disabilities is regulated by the bodies of executive power controlling the education sector. The functions of the Center can be referred to the existing or new educational institutions. The regulations of educational institutions must contain the corresponding statements on organization of the study process including that of the children with dishabilles (Mkrttchian et al.. 2015a)

The compulsory condition for organizing the study process with the use of distance learning technologies is the staff qualified accordingly and the rooms with the necessary equipment for managing distance learning technologies. The center can be created in educational institutions, special (correction) educational institutions for students with disabilities, children in need of psychological, medical and social support; also in institutions of higher education, regional resource centers for distance learning (Mkrttchian et al., 2015b).

The center can work via branches: the number of branches is determined in accordance with social, demographic, geographical and other features of the subject.

The center can pursue the following activities:

- Organizational and methodological support of distance learning for children with disabilities in the region;
- Providing the students' and instructors' access to the course packet necessary for realization of educational programs, and also to other electronic educational resources;
- Organization of the study process, help for students, instructors, parents (or their legal representatives);
- Collecting and analyzing data on the distance learning activity in the region.

Organization of the Study Process Based on ODL

The target of the study process includes children with disabilities studying at home in accordance with the educational programs of primary, secondary and higher education with the use of distance learning technologies. To this category refer students including children in need of special (correction) educational program: deaf, blind, visually impaired, hearing-impaired, with hard speech disorders, locomotive disorders, etc.

On the initial stage of the project ODL can be provided to the children without hard impairment's. With resources available, the study process can be organized in two groups: for children with health limitations and with children with hard defects. Composition of groups for ODL of children with disabilities is realized on agreement of their parents/ legal representatives with recommendations of the psychological, medical and pedagogical committee of the educational institution or in the individual program of a child's rehabilitation. It is recommended to realize ODL individually or in small groups. The number of students in groups can vary in accordance with the discipline. The general number of children with disabilities for participation in the initial stage of the project can be defined on the basis of the technical, organizational and personnel potential of the regional educational system.

The system created and functioning in this way can be used for home instruction of the children with disabilities with the view of their getting additional education. Instruction of the children with disabilities can be provided by teachers with the necessary knowledge of their physical and psychological development, and also with the skills of working with technical equipment. Courses of professional development for teachers could contribute to formation of their competence. The teachers should also be provided with necessary assistance and supervision (Mkrttcvhian, et al., 2015b).

CONCLUSION

Another illustration of the sliding mode functioning in the sphere of education was given in describing the terms of adaptive organization of the educational process of persons with disabilities with the use of open and distance learning technologies. High adaptability makes it possible to expand ODL for persons with disabilities, to transfer them to the new, more advanced living standards and create a new technology of instruction - open and distance lifelong learning (ODLL) technology. Adaptability (and, consequently, the quality of education) increases through the implementation of individual education programs. The use of intellectual agents in the sliding mode can provide for the quality individual trajectory of persons with disabilities.

REFERENCES

Mkrttchian, V., Aysmontas, B., Udin, A., Andreev, A., & Vorovchenko, N. (2015b). The Academic Views from Moscow Universities on the Future of DEE at Russia and Ukraine. In G. Eby & V. Yuser (Eds.), *Identification, Evaluation, and Perceptions of Distance Education Experts* (pp. 32–45). Hershey, PA: IGI Global. doi:10.4018/978-1-4666-8119-4.ch003

Mkrttchian, V., Bershabsky, A., Bozhday, A., & Fionova, L. (2015a). Model in SM of DEE Based on Service-Oriented Interactions at Dynamic Software Product Lines. In *Identification, Evaluation, and Perceptions of Distance Education Experts* (pp. 231-248). Hershey, PA: IGI Global.

KEY TERMS AND DEFINITIONS

Distance Education Expert (DEE): High level specialist in DE.

Feedback Control: Control System in Distant Education (DE).

Information System: is a system composed of people and computers that processes or interprets information.

Learning Environment: is combination of various educational technologies (including at least one communication module).

Pedagogical and Engineering Features: is effective application in education is not technological and managerial challenges of modern education, since its solution involves regulation of relations between all subsystems and elements of the educational system.

Student Avatar: Personalized graphic file or rendering that represents a computer user used to represent student in an online environment.

Teacher Avatar: Personalized graphic file or rendering that represents a computer user used to represent moderator in an online environment.

Chapter 6
Providing Quality Education for Persons With Disabilities Through the Implementation of Individual Educational Programs Managed by the Intelligent Agents in the Sliding Mode

ABSTRACT

Professional education of persons with disabilities is an important sphere of education enabling psychically and physiologically impaired persons to get economic independence contributing to their integration in society. The quality of professional education for persons with disabilities is realized only in conditions considering specifics of communicative and cognitive activity of the students with different disability categories. The absence of these conditions in universities makes it impossible for this category of students to complete the programs of higher education. As a rule, the contents of the study programs and the study schedule do not take this category of students into consideration.

DOI: 10.4018/978-1-5225-2292-8.ch006

INTRODUCTION

In unequal conditions such students have weak motivation for study, insufficient level of professional skills, they develop a consumer position. Apart from this, the reasons hindering the process of learning include architectural unavailability, insufficient psychological, medical and pedagogical support of study process; uneasiness of the teaching staff for instructing persons with disabilities.

Therefore, in conditions of unequal starting opportunities the students of this category cannot compare their success with the rest of the class and lose the strife for life goal. There arises the contradiction between the necessity of the providing education to persons with disabilities and inadequacy of the study process organization in university.

Individual study programs are based on humanistic principles, personality-oriented, differentiating, subject approach which makes them an effective mechanism of professional education for persons with disabilities.

MAIN FOCUS IN CHAPTER

Solutions and Recommendations

We suppose that in the process of implementer individual study programs in university education will provide the following effects:

- Individual opportunities of students with disabilities will be considered;
- Study process will become personality-oriented;
- Students with disabilities will be more successful;
- The quality of education will not be evaluated in categories of formal academic success but also include achievements in development of creative potential, formation of the wide range of competences and socialization of students;
- There will be a professional dialog between the specialists realizing support of persons with disabilities and instructors working on increasing their professional competences.

In accordance with the above let us turn to the definition of the quality of education given in normative documents. The quality of education highlights the two aspects of education:

1. Correspondence of goals and results of education to the modern social demands connected with transition to the open democratic society with market economy demanding making independent decisions on the basis of social experience, living in the situation of social and labor mobility, increasing tolerance level;
2. Correspondence of conditions of educational activity to the demands of students' health preservation and maintaining psychological comfort for all participants of the study process.

Apart from that, we consider education as a common process of education and development, the quality of education – as a quality of personality, its moral and civilian development. The quality of education is viewed as a social category which defines the state and outcomes of educational process in society, its correspondence to demands and expectations of various social groups in developing civil, common and professional competences of the students.

Individual study track is defined as a projected differentiated study program making the student a subject in the situation of choosing the program, participating in its development and realization, with the instructors' support of his/ her self-realization. Individual study track is determined by educational needs, abilities and opportunities of the student with disabilities, and also existing standards of education content. Individual study track is a structured program of the student's actions at a certain stage of study.

Individual study programs in the university educational space realize educational needs and opportunities of the students with disabilities due to the functions they perform:

- Fixing students' load, defining the order of following the study schedule and helping with the choice of the study track;
- Defining the goals, values and results of the students' educational activity;
- Allowing for realization of the students' needs in self-realization on the basis of choice, etc.

Analyzing the above, the quality of professional education for persons with disabilities can be defined as a condition of organizing educational activity corresponding to the demands of the students' health preservation and maintaining the psychological comfort of all participants of the educational process; organization of the study process based on introduction of individual educational programs, definition of individual study track in accordance with individual possibilities of persons with disabilities.

It is important to find the forms of study activity which could make the student feel successful and at the same time, could create conditions for his/her mastering the key competences.

Availability of higher professional education for persons with disabilities is maintained due to the three components: support, barrier-free environment and adaptive study process. The combination of these components enables the students with disabilities to successfully cover the educational standards and adapt themselves in society.

The indices of the high quality of professional education are:

- Availability for various categories of students;
- High demand in labor market;
- Correspondence of the education content to its goals and cognitive abilities of students with disabilities;
- Development of general cultural and professional competences of students;
- Formation of the skills necessary for acquiring knowledge in the course of life and information literacy.

Accordingly, the study process for persons with disabilities should be based on individual study programs for students with disabilities. Their use will provide for the high quality of education. Therefore, the development and construction of the individual study programs should follow the plan below:

- Defining the content of general educational programs;
- Developing and approving basis and working syllabi, and also the contents of the course programs, internships, final exams;
- Planning the contents of educational modules and academic load of the students and instructors;
- Appointing academic consultants for supporting the student with disabilities while planning and realizing the individual study track;

- Defining the methodological support of individually-oriented study process;
- Defining is the demands for material, technical and information support of the classes, etc.

Implementation of individual study programs in the process of learning for persons with disabilities should be backed up by the individual and differentiating approach. Differential of learning means considering individual and typological specificity of personality in the form of grouping the students and variability of the learning process in the groups. A differentiation of education highlights the introduction of certain changes in the course of study process for some groups of students. The term "differentiating approach" is defined as the approach to the process of learning based on differentiation in various kinds and forms.

Of special importance in the process of realization of individual educational programs is the choice of forms and methods of the study process organization. In classifying the methods of learning we followed our ideas (Mkrttchian et all, 2014; Mkrttchian, 2015) who singled out three groups of methods:

- Organizing the study activity;
- Stimulating the study activity;
- Controlling the efficiency of the study activity.

Most efficient are the methods of organizing the study activity of persons with disabilities based on the dialog communication of the students with disabilities, healthy students and the instructor with the view of mutual control, creation of situation of mutual participation, etc. Individuality of educational process is realized through various forms of working with the student: individual tasks, consultations, passing the test, organization of pair and group work, rehabilitation program, complex medical psychological and pedagogical support, etc. Implementation of individual education in quality education with both instructors and students involved should include the following stages:

- Revealing psychological and physiological specifics of the students with disabilities;
- Analyzing the results of the previous education of persons with disabilities;

- Developing methodological recommendations for education of persons with disabilities with selection of optimal instruction methods, the style of study interaction, the forms of knowledge control, etc.
- Constructing individual study programs on the basis of corrected syllabi in accordance with the needs and opportunities of the students;
- Advising the teaching staff realizing study programs on individual psycho-physical specificity of each student's development, its possibilities and limitations;
- Compiling a special schedule, individual load of the students; advising parents about their participation in the process;
- Implementing compulsory and voluntary assessment of the achievements of the persons with disabilities;
- Intermediate testing of students; searching for adequate means of control over covering the study program standards;
- Instructors' awareness of their activity being oriented to achievement of students' success, their development and socialization.
- Thus, implementation of the individual study track in the university educational process will result in the following changes:
- Increasing share of students successfully covering educational programs;
- Increasing number of psychologically protected students covering the programs according to their possibilities (extending the term of the programs, individual study schedule, etc.);
- Increasing share of successful students participating in compulsory and voluntary assessment of education quality.

CONCLUSION

Implementing individual programs in education process will result in satisfaction of education needs of students with disabilities which will allow for their passing to the new level of development, teach them to be responsible and independent in the future, to pursue their life goal.

REFERENCES

Mkrttchian, V. (2015, January-June). Use Online Multi-Cloud Platform Lab with Intellectual Agents: Avatars for Study of Knowledge Visualization & Probability Theory in Bioinformatics. *International Journal of Knowledge Discovery in Bioinformatics*, *5*(1), 11–23. doi:10.4018/IJKDB.2015010102

Mkrttchian, V., Kataev, M., Shih, T., Misra, K., & Fedotova, A. (2014, July-September). Avatars HHH Technology Education Cloud Platform on Sliding Mode Based Plug- Ontology as a Gateway to Improvement of Feedback Control Online Society. *International Journal of Information Communication Technologies and Human Development*, *6*(3), 13–31. doi:10.4018/ijicthd.2014070102

KEY TERMS AND DEFINITIONS

Control in E-Learning: Is the fundamental idea of control in learning is to reach the highest level of effectiveness in undertaking a task, which happens when individual's ability level is congruent with the level of challenge.

Didactic Method: Is a teaching method that follows a consistent scientific approach or educational style to engage the student's mind.

Learning Goals and Objectives: Is joint pedagogical aspects.

Online and Blended Learning of Adults: Is learners are adults, and training is carried out continuously throughout life. The purpose of adult education is closely associated with certain socio-psychological, occupational, household, personal problems, or factors or conditions with sufficiently clear ideas about further application of acquired knowledge, skills and qualities.

Studying and Training in Joint Activities: Is organizational aspects for training.

Virtual Assistant: Is special soft program creating in HHH University for training.

Chapter 7
Regulation of Discourse in Accordance With the Speech Regulation Factors Creating Conditions for Adaptability to the Situation

ABSTRACT

Realization of the above factors of speech regulation considered in this chapter, in our view, corresponds to the principle known as sliding mode. The sliding mode technique realized in communication as speech regulation principle provides for the flexibility of discourse, namely virtual political discourse and its adaptability to the communication situation as well as to the standards. Generally speaking, virtual political communication is regulated by the standards of diplomatic discourse and censorship less than classic political communication which is connected with the anonymity of online communicators.

DOI: 10.4018/978-1-5225-2292-8.ch007

INTRODUCTION

As it has been pointed out in Chapter 2, sliding mode in communication, particularly virtual communication, may be regarded as a technique of regulating discourse in accordance with the speech regulation factors creating conditions for adaptability to the situation (Aleshina, 2016).

The target content of speech (utterance) as the major regulation factor determining the outline of communication can be described in terms of dictemic information. The utterances-dictemes of political conflict communication can be marked by the atonality connected with the controversies of domineering ambitions. According to Sheigal, political discourse features an invariant speech act specificity (a set of speech acts) linked to the characteristics of a politician linguistic personality. The set of invariant speech acts reflects its functional semiotic triad "integration – orientation – atonality/ aggression" (Sheigal, 2000). Specific for conflict political discourse are the following speech acts: demand/ invocation, accusation/ verdict, irony/ sarcasm, warning/ threat. The information actualized in the utterance-dicteme (factual, intellectual, emotive, and impressive) provides for realization of speech act atonality of different degrees. We shall illustrate the above by the excerpt from the address to the nation by US president G.W. Bush dated March 17, 2003 (Bush, 2014). The address was devoted to the problem of Iraq and contains an ultimatum to the Iraqi leader Saddam Hussein. The address may be regarded both as classic and virtual communication and is characterized by atonality manifested in different parts of the text with a different degree. The following dicteme is marked by the high atonality level:

Saddam Hussein and his sons must leave Iraq within 48 hours. Their refusal to do so will result in military conflict, commenced at a time of our choosing. For their own safety, all foreign nationals - including journalists and inspectors - should leave Iraq immediately (Bush 2014).

This dicteme of orientation type possesses, mainly, communicative and orientation information. Meanwhile, this dicteme is quite impressive, its impressively being connected with the semantics of obligation expressed impressively in rather simple syntactic structures.

If Saddam Hussein attempts to cling to power, he will remain a deadly foe until the end. In desperation, he and terrorists groups might try to conduct terrorist operations against the American people and our friends. These attacks are not inevitable. They are, however, possible. (Bush, 2014)

This dicteme develops the idea expressed in the previous excerpt. The speaker clearly and straightforwardly outlines his intentions using relatively simple syntactic structures containing intellectual information.

MAIN FOCUS IN CHAPTER

Solutions and Recommendations

In the analyzed text we can also find the dictemes with a lesser degree of atonality. For example:

The United States, with other countries, will work to advance liberty and peace in that region. Our goal will not be achieved overnight, but it can come over time. The power and appeal of human liberty is felt in every life and every land. And the greatest power of freedom is to overcome hatred and violence, and turn the creative gifts of men and women to the pursuits of peace.

That is the future we choose. Free nations have a duty to defend our people by uniting against the violent. And tonight, as we have done before, America and our allies accept that responsibility (Bush, 2014)

The above dictemes realize speech acts of integration and orientation. Stated in the text is the US initiative against violence and hatred, rather a key pragmatic motive of the political conflict "us" – "them"/ "friend" – "enemy".

The speech under consideration, no matter how atonal it might be, is referred to the formal register of speech. The same seems true for the official online debate which is evident in the following debate excerpt between Australian federal election candidates Malcolm Turnbull and Bill Shorten.

Why is university becoming so unaffordable? Trent from Lindsay, NSW asks about quality education given it's crucial for the success of businesses, companies, individuals and the wider Australian community.

***PM:** It is vitally important that university is accessible to everyone. That's why we have HECS. We're seeking to reform and provide more flexibility to* *universities. We are not going to deregulate fees entirely. We will seek to offer universities the ability to deregulate fees ... for a small number of flagship courses so that they can compete.*

Hildebrand: *Doesn't that mean people with more money will get into top courses?*

PM: *I completely disagree with that.*

Shorten: *Labor doesn't support Mr. Turnbull's party cutting 20 per cent of university funding. We don't support the deregulation of university fees. We'll provide a minimum student funding guarantee of $10,800 per student per year. Hildebrand: Is it not true that we are only going to see the benefits of that in a generation's time ... at best?*

Shorten: *The idea that somehow you can be an innovation nation without being an education nation is political rubbish (Turnbull & Shorten, 2016).*

Whereas official political communication is ruled by regulations of diplomacy and political etiquette, online political discussions and blogs, especially those with anonymous posts, are often free from restrictions and censorship, thus possessing wider opportunities for being impressive and emotive. This may be illustrated by the excerpt from the online forum Debate Politics on 2016 US presidential elections. Agonality is connected with the conflict prompted by political rivalry of the candidates.

Originally Posted by El Veto-Voter Conservative Hillary is Dem; Progressive Trump is Rep; Really? Give us a break! Wow! What a wacky lineup! On the "LIBERAL" side, we have a long-time establishment politician who has supported just about everything that her followers oppose... On the "CONSERVATIVE" side, we have a long-time liberal donor who has supported just about everything that his followers oppose... If either were on a forum acting that way, they would immediately be called out for being a troll. (Debate Politics forum, 2016).

Most posts by the third party of the conflict (evidently, members of the electorate) are anonymous and are characterized by a high degree of agonality and are rather impressive. They vary from the unofficial to intensely unofficial register of speech.

Let us now refer to the second factor of regulation of speech communication, personal characteristics of the speaker which in a political conflict situation is traditionally determined by political communication norms. This factor also illustrates the acting sliding mode principle applied to communication. Psychologically, a politician as a speaker can pursue either a strategy of partnership or a strategy of assertiveness. These strategies are closely connected with the tactics of conflict behavior (unwillingness to see and recognize the differences, denying the conflict itself; concession marked by the aspiration to establish and improve relations by means of smoothening contradictions; confrontation (competition, rivalry) etc.). The second factor of regulation of speech communication is closely connected with the first one as the strategies can be traced in the content of the utterance.

It should be noted though that a politician striving for popularity ratings, on the general basis, tries to avoid compromising which may be regarded by the electorate/ nation as a weakness. The perfect example of assertive campaign speech may be one of Hillary Clinton 2016 campaign speeches (Clinton, 2016),

And to all of your supporters here and around the country: I want you to know, I've heard you. Your cause is our cause. Our country needs your ideas, energy, and passion. That's the only way we can turn our progressive platform into real change for America. We wrote it together -- now let's go out there and make it happen together. My friends, we've come to Philadelphia -- the birthplace of our nation -- because what happened in this city 240 years ago still has something to teach us today. We all know the story. But we usually focus on how it turned out -- and not enough on how close that story came to never being written at all when representatives from 13 unruly colonies met just down the road from here, some wanted to stick with the King. Some wanted to stick it to the king, and go their own way. The revolution hung in the balance. Then somehow they began listening to each other ... compromising, finding common purpose. And by the time they left Philadelphia, they had begun to see themselves as one nation. That's what made it possible to stand up to a King. That took courage. They had courage. Our Founders embraced the enduring truth that we are stronger together. America is once again at a moment of reckoning. Powerful forces are threatening to pull us apart. Bonds of trust and respect are fraying. And just as with our founders, there are no guarantees. It truly is up to us. We have to decide whether we all will work together so we all can rise together. Our country's motto is e pluribus Unum: "out of many, (we are) one (Clinton, 2016).

Our observations have shown that most of virtual political discourse on the part of the speaker is marked by a degree of assertiveness depending on the conflict situation proper as well as on personal psychological characteristics of the speaker. In case of anonymous communication, the degree of assertiveness increases substantially. We agree with I. Rowe, who states that when it comes to political discussion the democratizing potential of the Internet has been well documented and is clear for all to see. Online discussion platforms afford users a relatively high level anonymity, encouraging participants to express dissenting views without fear of retribution, and allowing previously disadvantaged and marginalized citizens to participate. "Despite its obvious potential, it is often argued that when it comes to discussing the types of divisive, emotional, and highly charged issues at the center of political debate, angry, hostile, and derogatory communicative behavior, as opposed to deliberation, has been one of the most widely recognized characteristics of online interaction. In fact, given the negative attention the Internet and its associated technologies have received in recent years, it may even be argued that online communication has become synonymous with uncivil communicative behavior. This is in large part thanks to the relatively high level of anonymity that is afforded Internet users when communicating online (Rowe, 2016).

The third factor of regulation of speech communication is the personal status of the listener/ reader. The success of communication depends much on his/ her personal background including education, social status and some personal psychological characteristics such as potential conflict behavior strategies. The reader/ listener of the political text may incorrectly perceive it because of the lack of expertise in politics which may result in negative, even aggressive reaction, as well. At the same time, anonymous online discussion may become an opportunity for a person to express his/ her views irrespective of the education level and background knowledge. Many online discussion platforms afford users a relatively high level anonymity, encouraging participants to express dissenting views without fear of retribution, and allowing previously disadvantaged and marginalized citizens to participate in discussions from which they may previously have been excluded. Such inclusive and diverse political discussion amongst citizens, it has been demonstrated, may provide a variety of positive democratic outcomes. Citizens who engage in political discussion will likely participate more in community affairs, be more tolerant of those who hold and/or express opposing political views, and make more informed and considered political decisions. Furthermore, citizens engaging in political discussion will likely feel more included in the democratic process, be better able to understand and justify their own preferences, and will set aside the adversarial approach to politics which often characterizes discussion on many controversial topics (Rowe, 2016).

The fourth factor of regulation of speech communication is the presence or absence of the persons who hear/ read the speech but are not involved in communication. This factor is closely connected with and determined by the fifth one – the properties of the communication link. Online political communication being realized in virtual space becomes visible and available for large audiences of readers. The fact that online political debates can be read from different parts of the world does not prevent the participants of discussions or comments authors from expressing their ideas quite openly and freely in terms of language register. Just the other way round, active communicators (generally anonymous) seem to enjoy the growing number of readers which may be connected with the above motives to express one.

The sixth factor, pre-supposition deals with the image the speaker/ writer has of the listener/ reader. This factor is closely connected with the seventh factor, the post-supposition, the assumption of the listener / reader about the personality of the speaker/ writer. Non-productive (post-suppositional unjustified) communication results from the discord in the ideas of communicators have of each other. These may lead to awkward situations and even prompt escalating of political conflicts. That was the case with the aforementioned speech made by G.W. Bush on Iraq dated March 17, 2003 (Bush, 2014). When addressing Iraqis, US president called for unity and moves for liberation. Below is the excerpt from the speech illustrating the call:

The day of your liberation is near. It is too late for Saddam Hussein to remain in power. It is not too late for the Iraq military to act with honor and protect your country, by permitting the peaceful entry of coalition forces to eliminate weapons of mass destruction. Our forces will give Iraqi military units clear instructions on actions they can take to avoid being attack and destroyed. I urge every member of the Iraqi military and intelligence services: If war comes, do not fight for a dying regime that is not worth your own life. And all Iraqi military and civilian personnel should listen carefully to this warning: In any conflict, your fate will depend on your actions. Do not destroy oil wells, a source of wealth that belongs to the Iraqi people. Do not obey any command to use weapons of mass destruction against anyone, including the Iraqi people. War crimes will be prosecuted, war criminals will be punished and it will be no defense to say, "I was just following orders." Should Saddam Hussein choose confrontation, the American people can know that every measure has been taken to avoid war and every measure will be taken to win it. Americans understand the costs of conflict because we have paid them in the past. War has no certainty except the certainty of sacrifice. Yet the only way to reduce the harm and duration of war is to apply the full force and might of our military, and we are prepared to do so. If Saddam Hussein attempts to cling to power, he will remain a deadly foe until the end (Bush, 2014).

Bush's constitutive rhetoric (Charland, 1991; Zagacki, 2007) did not correspond to the realities of mass consciousness in Iraq at the time and numerous domestic interethnic and inter-confessional controversies which made the call for liberation and following the example of American democracy senseless for many Iraq citizens.

CONCLUSION

Virtual political communication is less regulated by the standards of diplomatic discourse. Instant feedback allows the online communicator to correct the non-productive political communication which corresponds to the logic of the sliding mode.

REFERENCES

Aleshina, E. (2016). Structural, Information, and Regulation Aspects of Political Online and Classic Communication. In V. Mkrttchian, A. Bershadsky, A. Bozhday, M. Kataev, & S. Kataev (Eds.), *Handbook of research on estimation and control techniques in e-learning systems* (pp. 329–341). Hershey, PA: IGI Global. doi:10.4018/978-1-4666-9489-7.ch023

Bush, G. (2014). War Ultimatum Speech from the Cross Hall in the White House. *The Guardian.* Retrieved April 2, 2014 from http://www.theguardian.com/world/2003/mar/18/usa.iraq

Charland, M. (1991). Finding a horizon and telos: The challenge to critical rhetoric Quarterly. *Journal of Speech.*, *77*(1), 71–74. doi:10.1080/00335639109383944

Clinton, H. (2016). *Democratic Presidential Nomination Acceptance.* Retrieved from http://www.americanrhetoric.com/speeches/convention2016/hillaryclinton2016dnc.htm

Debate Politics Forum. (2016). Retrieved from http://www.debatepolitics.com/2016-us-presidential-election/263177-troll-presidential-candidates.htm

Rowe, I. (2016). *Online political discussions tend to be less civil when the participants are anonymous*. Retrieved from http://www.democraticaudit.com/?p=1455

Turnbull, M., & Shorten, B. (2016). Retrieved from http://www.news.com.au/national/federal-election/malcolm-turnbull-and-bill-shorten-faceoff-in-australias-first-online-leaders-debate/news-story/1b34234a057303744a32ef6a1c8a5e57

Zagacki, K. S. (2007). Constitutive rhetoric reconsidered: Constitutive paradoxes in G.W. Bushs Iraq war speeches. *Western Journal of Communication*, *71*(4), 272–293. doi:10.1080/10570310701653786

Chapter 8
Complex Social, Medical, Psychological, and Educational Support for People with Disability Act (IDEA)

ABSTRACT

For young people with disabilities the start of university study is equal to the first step towards social integration as their previous stages of education took place in institutions of segregation type. It happens in the situation of crisis of transition from educational space to another. Meanwhile, institutions of primary, secondary and higher education reflect the acting model the psychological and pedagogical integration of persons with disabilities in society. Inclusion of persons with disabilities is regarded as a stage of their getting social adaptation and integration.

INTRODUCTION

The importance of the problem substantially increased in 2008 due to the Russian Federation joining Convention of the rights of persons with disabilities declaring the right and possibility of the children, youth and adults with disabilities to live and study in the space with minimum limitations aimed at their social integration.

DOI: 10.4018/978-1-5225-2292-8.ch008

From the economic and social perspective, it is important to guarantee persons with disabilities the possibility for building professional future, integration in the sphere of manufacture and society. We provide persons with disabilities with full education; we will be able to pass from the distribution concept and social support to rehabilitation concept. The fully educated person with disabilities has more chances to get a qualified job and appropriate position life status.

Scholars give much importance to developing the question of transition from school education to professional activity of the youth with light physical and psychic disorders, those who are potentially capable of active labor activity, career growth, economic self-support and social self-realization. One of the conditions providing for academic, social and personal success of persons with disabilities in institutions of professional education is a complex psychological, pedagogical, medical and social support as an integral part of educational process.

Despite the actual character of researching into this problem, the scholars highlight serious blanks in its scientific background and the absence of adequate theoretical and methodological approaches to the problems of the development of persons with disabilities as subjects of professional and educational activity. Of high importance are various aspects of supporting professional education of persons with disabilities both in the situation of their learning in homogeneous groups and in heterogeneous (mixed) ones. To be studied are the goals, tasks and forms, contents and methods of supporting professional education in various types of institutions. Moreover, the above is to be presented as a system backed up scientifically and methodologically (Mkrttchian, 2015).

Working out issues of complex support of professional education for persons with disabilities is based on the understanding of the term "support" in contrast to the terms "help" and "support" as well as with determining its goals, tasks and designing its structural and functional model.

Support requires immediate interaction and contact of the teacher and the student unlike the assistance and help which can be provided from the distance. Support means certain actions by the instructor, while assistance and help can have the form of recommendations for the student to realize the necessary steps. Support is based on the results of diagnostics and requires projecting the undertake actions, whereas assistance and help are of operational character and can be performed according to the instructor's intuition and experience.

Finally, support is a more wide scale pedagogical phenomenon including both help, and assistance.

In the context of complex tasks to be solved by the education system in social, demographic and psychological conditions of the start of the 21st century, support of education process is often regarded as the phenomenon complex in its structure, the terms “psychological-pedagogical, medical-social support”, “psychological, medical and pedagogical support”, “complex support” have entered the scientific vocabulary to denote a special kind of assistance to the child, his/her parents, instructors in solving complex tasks connected with education, medical treatment, socialization of the maturing person as personality. For example, psychological, medical and pedagogical support offers a wide range of long-term measures of complex help realized in the process of agreed work of various specialists (teachers, dialectologists, medical workers, social teachers) (Mkrttchian, 2016).

MAIN FOCUS IN CHAPTER

Solutions and Recommendations

Teaching persons with disabilities in institutions of primary and secondary professional education supposes organization of complex psychological-pedagogical, medical-social support aimed at, on the one hand, creating conditions adequate to their individual needs, on the other hand, at preventing situations and risks of adaptation violations in social and personal development.

Both aspects of complex support seem equal. The first one determines academic success of persons with disabilities and gives them the possibility to master professional competences at sufficient level.

The second aspect is linked with providing conditions for development of social competences and personality adaptation potential allowing for active adaptation to the changing environment with help of various social means. Systemic character of the goals set shows that such activity should be performed by the team including pedagogical workers of educational institutions as well as pedagogical workers of “supporting” professions – special instructors (dialectologists), special psychologists, social teachers, doctors, tutors, etc.

Explication of essential characteristics of pedagogical support in the wide educational context revealed the following ones:

- Activity nature of support supposing the influenced exercised on the supported phenomenon;
- Control connected with a certain given optimal trajectory defining the development of the supported phenomenon;
- Individual character (content targeting corresponding to the conditions and specificity of the supported process);
- Functioning in specially created space setting optimal conditions for existence of the supported phenomenon;
- Continuity of realization, beginning and end of support;
- Consideration of monitoring results which determine the contents of support actions.

The efficiency of the support service will be determined by following the conditions listed as follows:

- Keeping the invariant algorithm of realizing complex support including diagnostic-analytical, projective, control and assessment stages;
- Individualization of tasks, directions, methods and technologies of support of professional education in accordance with the needs of persons with disabilities and the social situations;
- Targeting the work of the support system at creating conditions providing for its academic, social and personal success of persons with disabilities and their social integration;
- Positive motivation of support service specialists for solving tasks of complex support.

CONCLUSION

Thus, organization of complex (psychological-pedagogical, medical-social) support as a special systemic activity aimed at providing conditions for professional education, development adequate to individual needs of persons with disabilities and prevention of adaptation violations risks in their physical, social and personal development will allow for expanding availability and improving the quality of education quality in institutions of various types. The form of institutionalization of support and its structural and functional model vary according to the administration of educational institutions and the needs of actual students and prospective ones as well as material resources of the institution.

REFERENCES

Mkrttchian, V. (2016). The Control of Didactics of Online Training of Teachers in HHH University and Cooperation with the Ministry of Diaspora of Armenia. In V. Mkrttchian, A. Bershadsky, A. Bozhday, M. Kataev, & S. Kataev (Eds.), *Handbook of research on estimation and control techniques in e-learning systems* (pp. 311–322). Hershey, PA: IGI Global. doi:10.4018/978-1-4666-9489-7.ch021

Mkrttchian, V., Kataev, M., Bershadsky, A., & Volchikhin, V. (2015). Use Triple H-AVATAR Technology for Research in Online Multi-Cloud Platform Lab. In A. Kravets et al. (Eds.), *Proceedings of CIT&DS 2015* (pp. 58–67). Springer International Publishing.

KEY TERMS AND DEFINITIONS

Functional Modelling Software Platform: A specification software designed to be used modelling of the risk management process of Enterprise Resource Planning on lab Multi-Cloud Platform has allowed us to solve the problem of compliance, as well as to identify modern and future issues, concepts, trends and solutions IS&T throughout the software life cycle.

Online Multi-Cloud Platform Lab: Laboratory on the Internet, which is available on the multi cloud platform and intended for research, training and development of forecasting.

Chapter 9
Tolerance as Reflection of Sliding Mode in Psychology

ABSTRACT

As it has been pointed out in Chapter 3, sliding mode in communication, particularly real communication, may be regarded as a technique of regulating tolerance problems. The problem of tolerance has not been fully covered despite the interest for it in various fields of science. With the accumulation of substantial empirical material, evident is the lack of generalizing research. Among scarcely covered are administration, political and professional tolerance.

INTRODUCTION

Despite the great number of works on the essence of tolerance, its structure and components are still to be described. There is also variety of views on definition of tolerance. Depending on the context, tolerance is filled with a specific sense. Most researchers state that psychological contents of tolerance cannot be concentrated on one property or characteristic due to the complexity of the phenomenon.

Integrity of tolerance was highlighted as early as the 1950-s. It was considered that this feature resulted from a number of forces acting in a personality in one direction: of temperament, social space and education in the family, acquired experience, sociocultural factors. It means the impossibility of describing tolerance as based on one concept (integrity) and in one dimension

DOI: 10.4018/978-1-5225-2292-8.ch009

only. The term “tolerance” should be regarded in two psychological senses: psychophysiological (resistance towards adverse factors of activity) and socio-psychological (tolerance to others, social tolerance).

The psycho-physiological aspect of considering tolerance links it to the measure of resistance to stresses, adverse factors of environment and own irritants. Literary sources on this problem prove that tolerance is connected with the main properties of the nervous system. Among psycho-physiological parameters backing up the productivity of activity and specifics of tolerant behavior in strenuous conditions, of great importance is the force of irritation process, mobility and balance of nervous processes. The trainings in free and forced sliding modes in both psychological senses have shown that the role of nervous system specifics in determination of a person’s tolerant behavior is not absolute. In some activities characterized by high emotion of conditions, either one, or the other of the two contrary poles of the parameter for each nervous system property can play both a positive and a negative role.

Tolerance as sustainability and resistance allows for adjustment to adverse factors, in this case it is characterized by adaptability.

In social aspect tolerance is regarded as a social and psychological phenomenon that manifests itself through the way of interacting with the outer space. In this sense, tolerance is expressed in strife for understanding and agreement in the process of communication and interaction by means of cooperation and dialog based on a person’s ability to accept another person in his/ her diversity.

For a modern professional, developing tolerance is turning into a strategically important goal. Tolerance is vitally important for any specialist and has certain specifics in various professions. Professional tolerance is one of the competence bases for a number of jobs as it determines economic efficiency or inefficiency of professional activity.

Tolerance may be regarded as integral personal characteristics of an individual defang its ability to interact with the outer world actively in problem situations. Tolerant personality possesses a set of qualities providing for its constructive professional development, having skills of tolerant interaction.

Tolerant interaction is characterized by the following manifestations: preference to kindness values, independence, responsibility, absence of expressed assessment, empathy, cooperation, activity, prevalence of active strategies of coping with difficulties, reconstructing the system of self-regulation in accordance with changing outer and inner conditions, high degree of reflexivity, internal control locus, optimism, emotional balance, ability to control impulsive behavior, responsibility, adequacy of self-assessment, positive self-perception, extraversion, openness for the new experience (Semionova, 2016).

MAIN FOCUS IN CHAPTER

Solutions and Recommendations

Professional tolerance is an integral personal quality serving the specialist's basis for resistance to adverse phenomena of professional development hindering the rise of professional destructions. This quality of a personality is a necessary ground for the constructive professional development.

The research of tolerance in sliding mode can follow the indices listed below (Mkrttchian et al., 2017):

- General index of personality tolerance according to the express-questionnaire "Tolerance index"
- Individual degree of expressed ability for self-analysis and understanding own condition, ability for understanding other people's actions;
- Degree of developed sense of personal responsibility for the events happening to the person (the level of internality);
- Five basic personal characteristics: neurotic, extraversion, openness to experience, cooperation according to the questionnaire of a professional's personal features diagnostics;
- Personality empathy according to the Questionnaire for diagnostics of empathy ability;
- Level of developed general self-regulation according to the questionnaire "Behavioral self-regulation style";
- Emotional burnout (the symptoms of professional deterioration, depersonalization, reduction of professional achievements according to the Questionnaire of psychic burnout;
- Coping behavior in stress situations of cooperation (task-oriented coping, emotion–oriented coping, avoidance coping according to "Coping-behavior in stressful situations").

The research may make use of the following techniques:

1. The questionnaire of professional burnout (consists of the scales reflecting emotional deterioration, depersonalization, reduction of professional achievements
2. Technique "Coping behavior in stressful situations"
3. The questionnaire for diagnostics of empathy abilities
4. The questionnaire "Behavioral self-regulation style"
5. Express-questionnaire "Tolerance index"
6. Questionnaire of subjective level control
7. Questionnaire of a specialist's personal qualities diagnostics

Of primary importance for a personality's constructive development is its own activity. Prevention of professional deformations may be expressed in the reflection of professional development and searching for ways of further personal professional growth. If a specialist remains within the system "person-profession: and does not pass to the level of reflation of his/her professional development in terms "person-world", his/her professional track will be characterized as destructive.

The scope of professional development problems is much wider than those mentioned above. Some of them have just been indicated, others are being solved. For instance, the contents of certain concepts characterizing professional development need clarifying. There arises a necessity to study the problems aimed at defining the specificity of a specialist's professional development during his/her university studies and at the stage of his/her realization as professional in various types of professions at different life stages. Covering the above problems is of high priority in the sphere of modern psychology.

Due to the fact that an illness of psychogenic character results from the patient's wrong decisions, it may be regarded as a result of his/her wrong decisions as a consequence of misconceptions, logical mistakes while analyzing psych traumatic situations and working out own actions. Accordingly, the psychotherapist's tasks include the following:

- Revealing misconceptions about the cause and development of the illness;
- Strengthening the patient's ability for logical argumentation with the view of creating the correct inner picture of the illness;
- Bringing awareness that the reason of the illness is connected with the insufficient use of rational opportunities of psychic activity which can help cope with the problems.

The basis for rational psychotherapy is logical argumentation. Apart from that the method includes explanation, suggestion, emotional influence, study and correction of personality, didactic and rhetorical techniques (Semionova, 2016).

The process of rational therapy requires developing a person's sensible and adequate attitude to the problem. Thus, the core of rational psychotherapy is the correct and comprehensible description of the illness causes and its prognosis. In its turn, it contributes to formation of the adequate attitude to the problem.

Let us consider the theoretical premises of the rational and emotional therapy.

The first premise of the rational and emotional therapy states the three spheres of a person's functioning: thinking, emotions and behavior which are linked to each other. Changing one of them influences the functioning of others. While behavioral therapy poses the task of reaching changes by affecting a person's environment, rational and emotional therapy is aimed at changing emotions by affecting the thinking contents. The possibility of such changes is based on connected thinking and emotions. From the viewpoint of rational and emotional therapy, cognitions are a major factor determining the emotional state. The individual's interpretation of an event results in the emotion he/she gets in this situation. Negative feelings are caused rather by our ideas about these events than the people and events themselves. Influence of thinking is the shortest way of reaching changes in our emotions as, consequently, our behavior.

The second premise serving as a base for rational psychotherapy deals with the questions of emotions pathology. From the viewpoint of rational therapy, pathological violations of emotions are based on the aberration of thinking processes and cognitive errors. For denoting all categories of cognitive errors we suggest using the term "irrational propositions". To these refer such forms of errors as exaggeration, simplification, ungrounded assumptions, erroneous conclusions. The rise of irrational propositions is linked to the patient's past when he/she as a child perceived the world still unable to critically analyze it, not having the possibility to contradict it on the behavioral level.

As rational and emotional therapy links pathological emotional reactions to irrational propositions, it is the fastest way to change the distress state by means of changing erroneous propositions. The mechanisms supporting irrational propositions usually exist in the present tense. Therefore, the rational and emotional theory does not concentrate on analyzing the past causes leading to formation of an erroneous assumption, but rather on analyzing the present. Rational and emotional theory finds out how an individual retains his/ her symptoms sticking to certain irrational convictions, why he/she does not refuse from them or correct them. The research has shown that irrational cognitions can be changed. Only the reconstruction of erroneous cognitions leads to changes in emotional reactions. During rational and emotional therapy a person acquires the ability to control his/her erroneous propositions contrary to the primary stage of the therapy when irrational propositions control the person's behavior.

Formal logic offers such concepts as incorrect and correct thinking. The thinking that can lead a person to correct solutions of theoretical and practical tasks is the correct thinking. It should satisfy three basis requirements: definiteness, consistency and argumentativeness. Certain thinking supposes accuracy, it is free from rambling. Consistency supposes the freedom of thinking from inner contradictions as well as logical links between certain ideas. Finally, argumentative thinking does not only forms the truth, but also shows the ground for the idea to be accepted as a true one, so it provides arguments for it.

The incorrect thinking, on the contrary, is characterized by inconsistence, absence of proof and indefiniteness. The inconsistency of thinking is most often based on violation of the law of equation expressed in the formula «A=A». In communication it happens when two arguing speakers understand one the same term differently. The terms "neurosis", "alcoholism", etc. may serve examples as the psychotherapist treats them differently from the patient for whom they convey only the common meaning.

The law of inconsistency of thinking was formulated by Aristotle. He stated that two contrastive ideas about one and the same object cannot be simultaneously truthful. For instance, it is impossible to find truthful the following ideas of an alcoholic: "I drink because I'm sick".

Such conclusions in formal logic are called analogisms. Sometimes analogisms arise when a patient considers the two propositions referring to different subjects as contradicting and excluding the possibility of the third one ("I'm neurotic because I have bad terms with colleagues at work, and at work it's bad because I'm neurotic"). These ideas exclude the third variant – another job, other colleagues, and other relationship.

Absence of argumentation is most often evident as a logical error: «after this means as a result of this». («I stammer because I was frightened by a dog»; «I'm afraid of heights because I fell down from the rock», etc.). Absence of proof in the patient's propositions is revealed when they originate from wrong grounds («My disease can be treated by hypnosis only, but I wasn't prescribed the hypnosis, so...»). The knowledge of equation law, ability to correctly operate the concepts helps the psychiatrist to make the thinking definite. The knowledge of contradiction laws, the skill of argumenting propositions well facilitates consistency of thinking. Finally, the knowledge of sufficient ground and the rules of deductive and inductive conclusions facilitate learning argumentative thinking. Thinking without argumentation can be contrasted only to the thinking based on logic and arguments. The psychotherapist can use any arguments if they are demonstrative and can lead to the patient's healing.

From the above it may be concluded that an important part of rational psychotherapy is correction of inconsistency фтв contradictions.

In the process of rational therapy the autogenetic analysis facilitates searching for the actual conflict of the personality with the environment.

The important principle addition to the above is the fact that a great number of a person's convictions are unconscious which means an additional stage in psych correction work aimed at realizing the person's key beliefs. Psychological methods, especially projective ones, can be of great help in this process.

Thoughts, convictions, inner dialog exercise creative influence on the person and his/her life script. They reveal themselves not only in behavior and feelings, but also in the readiness to overcome life stresses.

Irrational thoughts and convictions serve the basis for "unhealthy" models of behavior. They are characterized by strong unpleasant emotions (anger, helplessness, annoyance). They are often pessimistic and suppose the unhappy outcome "I won't be able to cope with that", "I won'e do it", etc.). They are based on helplessness.

It is useful to eliminate all destructive expressions from speech, self-addresses with self-accusations and absence of belief in success ("I won't be able to change my life", "Nobody understands me", etc.). It is important to start thinking in a constructive way not to stick to negative emotions and not to hinder the process of coping with life difficulties and professional stresses.

Positive thinking in sliding mode is the ability to accept the situation without evaluating it. It would be useful to repeat positive statements for positive self-suggestion.

CONCLUSION

Sliding mode finds its application in psychology and psychotherapy opening new opportunities in correcting professional burnout and neurotic states. The phenomenon of tolerance which is quite actual today, reflects the logic of the sliding mode and can be studied and described accordingly

FUTURE RESEARCH DIRECTIONS

Future research may focus on the application of results of emerging research of intellectual control in sliding mode accompanied by the integration of different components of measuring facilities, computing hardware, comparison devices and actuation devices with a view of creating new generation systems - Embedded Systems & Networks for Cyber Control and Communication (Triple C) of Cloud computing in Sliding Mode. In this case, we would expect interoperability to be a big factor in the next decade. Cloud technology providers are still a bit skeptical of the concept but eventually. We hope that technology will move in this direction and allow Cloud applications to be platform agnostic. This means Cloud developers will be free of risk in getting locked-in to a platform. It is not necessarily good news for Cloud owners but at the same time, Clouds will be able to share resources and form federations in order to balance their loads more effectively and therefore increase their profits. Various auction mechanisms are being discussed taking into consideration factors such as the energy consumption of a Cloud, the QoS it can deliver and the amount of renewable resources it is using for power. These mechanisms are then used by service providers to determine which Cloud is better suited for their application and also allow the dynamic reallocation of services when criteria and conditions change. This is where we believe Cloud technology can make a difference in traffic management since we can create mechanisms that take into account the QoS required by an application and the network locations of its clients and find which Cloud is better suited for hosting it. As more players enter the Cloud market and more people are willing to use the technology, we will start seeing many datacenters appearing in different locations worldwide. In the future it could make sense to look for ways of federating all these datacenters and operate them as a big "Cloud of Clouds in Sliding Mode for Intellectual Control and Communication for non –engineering system in Humanitarian and social spheres".

REFERENCES

Mkrttchian, V., Amirov, D., & Belyanina, L. (2017). Optimizing an Online Learning Course Using Automatic Curating in Sliding Mode. In N. Ostashewski, J. Howell, & M. Cleveland-Innes (Eds.), *Optimizing K-12 Education through Online and Blended Learning* (pp. 213–224). Hershey, PA: IGI Global. doi:10.4018/978-1-5225-0507-5.ch011

Semionova, E. (2016). Professional deformation of representative's sociologist professions at different stages of professionalization. *Psychotherapy in Practice, 58*(1), 88-94.

Glossary

Chattering: It is results in chattering, an undesirable phenomenon of applying SMC, where high frequency switching is applied to the system during the sliding phase.

Construction of Equivalent Control Action: Is the system motion along the sliding surface can be interpreted as an *average* of the system's dynamics on both sides of the sliding surface.

Control in E-Learning: Is the fundamental idea of control in learning is to reach the highest level of effectiveness in undertaking a task, which happens when individual's ability level is congruent with the level of challenge.

Determination of Equivalent Control: Is the system motion along the sliding surface can be interpreted as an *average* of the system's dynamics on both sides of the sliding surface.

Didactic Method: Is a teaching method that follows a consistent scientific approach or educational style to engage the student's mind.

Discontinuous Control: It this brings the discontinuity to the control, and the whole nonlinear system.

Distance Education Expert (DEE): High level specialist in DE.

Education Technology: Are technical, biological and engineering systems for Education whose components are combined, controlled and generated using the aligned single processing core. All the components at all levels of interaction are combined in the network infrastructure. All components include built: in calculators, providing data processing in real: time.

Feedback Control: Control System in DE.

Functional Modeling Software Platform: A specification software designed to be used modeling of the risk management process of Enterprise Resource Planning on lab Multi: Cloud Platform has allowed us to solve the problem of compliance, as well as to identify modern and future issues, concepts, trends and solutions IS&T throughout the software life cycle.

Indicator of Sliding Mode: The software for control virtual research space, maintain it sliding mode.

Information System: Is a system composed of people and computers that processes or interprets information.

Learning Environment: Is combination of various educational technologies (including at least one communication module).

Learning Goals and Objectives: Is joint pedagogical aspects.

Moderator Avatar: Personalized graphic file or rendering that represents a computer user used to represent moderator in an online environment.

Online and Blended Learning of Adults: Is learners are adults, and training is carried out continuously throughout life. The purpose of adult education is closely associated with certain socio: psychological, occupational, household, personal problems, or factors or conditions with sufficiently clear ideas about further application of acquired knowledge, skills and qualities.

Online Multi-Cloud Platform Lab: Laboratory on the Internet, which is available on the multi cloud platform and intended for research, training and development of forecasting.

Pedagogical and Engineering Features: Is effective application in education is not technological and managerial challenges of modern education, since its solution involves regulation of relations between all subsystems and elements of the educational system.

Researcher Avatar: Personalized graphic file or rendering that represents a computer user used to represent researcher in an online environment.

Second Order Sliding Mode (SOSM): Is order reduction phenomenon was reviewed above and partial dynamic collapse is the reduction of order for the compensated dynamics of the SMC of the system.

Sliding Mode Control (SMC): Is control algorithm for adjustment of learning tasks.

Student Avatar: Personalized graphic file or rendering that represents a computer user used to represent student in an online environment.

Studying and Training in Joint Activities: Is organizational aspects for training.

Teacher Avatar: Personalized graphic file or rendering that represents a computer user used to represent moderator in an online environment.

Tolerance: Is an integral personal quality of an individual providing for its ability to actively and positively interact with the out world in problem situations. It functions as an important component of the life attitudes of a person possessing of own values and interests are able to protect them, if necessary, at the same time treating others' positions and values with respect.

Tolerant Personality: Is a personality possessing of a certain set of qualities providing for its constructive professional development characterized by skills of tolerant interaction. Tolerant personality is characterized by the following manifestations: value preference of kindness, independence, responsibility, absence of marked evaluability, empathy, cooperation, activity, prevalence of active strategies of coping with difficulties, the ability to flexibly reconstruct the system of self: regulation as prompted by changing outer and inner conditions, high degree of reflexivity, manifestation of the internal control locus, optimism, emotional stability, ability to control impulsive behavior, responsibility, appropriateness of self: assessment and positive self: perception, extraversion, openness to the new experience.

Triple H-AVATAR Technology: The technology of modeling and simulation based on known technology of Avatar used in the HHH University since 2010.

Virtual Assistant: Is special soft program creating in HHH University for training.

Virtual Research Environment: The space where with the help of virtual reality creates a special environment for research.

Related Readings

To continue IGI Global's long-standing tradition of advancing innovation through emerging research, please find below a compiled list of recommended IGI Global book chapters and journal articles in the areas of sliding mode, educational applications, and virtual communications. These related readings will provide additional information and guidance to further enrich your knowledge and assist you with your own research.

Abbassi, R., Saidi, S., Hammami, M., & Chebbi, S. (2015). Analysis of Renewable Energy Power Systems: Reliability and Flexibility during Unbalanced Network Fault. In A. Azar & S. Vaidyanathan (Eds.), *Handbook of Research on Advanced Intelligent Control Engineering and Automation* (pp. 651–686). Hershey, PA: IGI Global. doi:10.4018/978-1-4666-7248-2.ch024

Abdel Aziz, M. S., Elsamahy, M., Hassan, M. A., & Bendary, F. M. (2017). Enhancement of Turbo-Generators Phase Backup Protection Using Adaptive Neuro Fuzzy Inference System. [IJSDA]. *International Journal of System Dynamics Applications*, *6*(1), 58–76. doi:10.4018/IJSDA.2017010104

Abdelmalek, S., & Belmili, H. (2015). A New Robust H∞ Control Power. In A. Azar & S. Vaidyanathan (Eds.), *Handbook of Research on Advanced Intelligent Control Engineering and Automation* (pp. 601–623). Hershey, PA: IGI Global. doi:10.4018/978-1-4666-7248-2.ch022

Ahrens, A., & Zaščerinska, J. (2015). A Comparative Study of Business and Engineering Students' Attitude to Mobile Technologies in Distance Learning. In P. Ordóñez de Pablos, R. Tennyson, & M. Lytras (Eds.), *Assessing the Role of Mobile Technologies and Distance Learning in Higher Education* (pp. 29–59). Hershey, PA: IGI Global. doi:10.4018/978-1-4666-7316-8.ch002

Akkarapatty, N., Muralidharan, A., Raj, N. S., & P., V. (2017). Dimensionality Reduction Techniques for Text Mining. In V. Bhatnagar (Ed.), *Collaborative Filtering Using Data Mining and Analysis* (pp. 49-72). Hershey, PA: IGI Global. doi:10.4018/978-1-5225-0489-4.ch003

Albu, F., & Nishikawa, K. (2015). Nonlinear Adaptive Filtering with a Family of Kernel Affine Projection Algorithms. In A. Azar & S. Vaidyanathan (Eds.), *Handbook of Research on Advanced Intelligent Control Engineering and Automation* (pp. 61–83). Hershey, PA: IGI Global. doi:10.4018/978-1-4666-7248-2.ch002

Almajano, P., Lopez-Sanchez, M., Rodriguez, I., Puig, A., Llorente, M. S., & Ribera, M. (2016). Training Infrastructure to Participate in Real Life Institutions: Learning through Virtual Worlds. In F. Neto, R. de Souza, & A. Gomes (Eds.), *Handbook of Research on 3-D Virtual Environments and Hypermedia for Ubiquitous Learning* (pp. 192–219). Hershey, PA: IGI Global. doi:10.4018/978-1-5225-0125-1.ch008

Andreu, L., & Sanz-Torrent, M. (2017). The Visual World Paradigm in Children with Spoken Language Disorders. In C. Was, F. Sansosti, & B. Morris (Eds.), *Eye-Tracking Technology Applications in Educational Research* (pp. 262–282). Hershey, PA: IGI Global. doi:10.4018/978-1-5225-1005-5.ch013

Aymen, F., Kraiem, H., & Lassaâd, S. (2015). Electrical Motor Parameters Estimator Improved by a Computational Algorithm. In A. Azar & S. Vaidyanathan (Eds.), *Handbook of Research on Advanced Intelligent Control Engineering and Automation* (pp. 567–600). Hershey, PA: IGI Global. doi:10.4018/978-1-4666-7248-2.ch021

Azar, A. T., & Serrano, F. E. (2015). Stabilization and Control of Mechanical Systems with Backlash. In A. Azar & S. Vaidyanathan (Eds.), *Handbook of Research on Advanced Intelligent Control Engineering and Automation* (pp. 1–60). Hershey, PA: IGI Global. doi:10.4018/978-1-4666-7248-2.ch001

Azar, A. T., & Serrano, F. E. (2016). Stabilization of Mechanical Systems with Backlash by PI Loop Shaping. *International Journal of System Dynamics Applications*, *5*(3), 21–46. doi:10.4018/IJSDA.2016070102

Bartoszewicz, A., & Zuk, J. (2010). Sliding mode control — Basic concepts and current trends. *2010 IEEE International Symposium on Industrial Electronics*, (1), 3772–3777. http://doi.org/ doi:10.1109/ISIE.2010.5637990

Bedoui, H., Kedher, A., & Ben Othman, K. (2015). Fault Detection and Isolation for an Uncertain Takagi-Sugeno Fuzzy System using the Interval Approach. In A. Azar & S. Vaidyanathan (Eds.), *Handbook of Research on Advanced Intelligent Control Engineering and Automation* (pp. 364–389). Hershey, PA: IGI Global. doi:10.4018/978-1-4666-7248-2.ch013

Behera, A. K., & Bandyopadhyay, B. (2015). *Self-triggering-based sliding-mode control for linear systems*. Academic Press. 10.1049/iet-cta.2015.0342

Ben Hariz, M., Bouani, F., & Ksouri, M. (2015). Design of a Controller with Time Response Specifications on STM32 Microcontroller. In A. Azar & S. Vaidyanathan (Eds.), *Handbook of Research on Advanced Intelligent Control Engineering and Automation* (pp. 624–650). Hershey, PA: IGI Global. doi:10.4018/978-1-4666-7248-2.ch023

Benahdouga, S., Boukhetala, D., & Boudjema, F. (2012). Electrical Power and Energy Systems Decentralized high order sliding mode control of multimachine power systems. *International Journal of Electrical Power & Energy Systems*, *43*(1), 1081–1086. doi:10.1016/j.ijepes.2012.06.018

Bennett, J., & Lin, F. (2017). iPad Usage and Appropriate Applications: K-12 Classroom with a 1-to-1 iPad Initiative. In N. Ostashewski, J. Howell, & M. Cleveland-Innes (Eds.), Optimizing K-12 Education through Online and Blended Learning (pp. 185-212). Hershey, PA: IGI Global. doi:10.4018/978-1-5225-0507-5.ch010

Biba, M., Vajjhala, N. R., & Nishani, L. (2017). Visual Data Mining for Collaborative Filtering: A State-of-the-Art Survey. In V. Bhatnagar (Ed.), *Collaborative Filtering Using Data Mining and Analysis* (pp. 217–235). Hershey, PA: IGI Global. doi:10.4018/978-1-5225-0489-4.ch012

Blomgren, C. (2017). Current Trends and Perspectives in the K-12 Canadian Blended and Online Classroom. In N. Ostashewski, J. Howell, & M. Cleveland-Innes (Eds.), *Optimizing K-12 Education through Online and Blended Learning* (pp. 74–92). Hershey, PA: IGI Global. doi:10.4018/978-1-5225-0507-5.ch004

Blomme, R. J. (2015). Internet Technology and its Application in Competence Development of Highly Educated Staff: The Role of Transfer. In P. Ordóñez de Pablos, R. Tennyson, & M. Lytras (Eds.), *Assessing the Role of Mobile Technologies and Distance Learning in Higher Education* (pp. 249–271). Hershey, PA: IGI Global. doi:10.4018/978-1-4666-7316-8.ch011

Blomme, R. J. (2015). The Role of Internet Technology in Higher Education: A Complex Responsive Systems Paradigm. In P. Ordóñez de Pablos, R. Tennyson, & M. Lytras (Eds.), *Assessing the Role of Mobile Technologies and Distance Learning in Higher Education* (pp. 228–248). Hershey, PA: IGI Global. doi:10.4018/978-1-4666-7316-8.ch010

Boga, S. R., Kansagara, B., & Kannan, R. (2017). Integration of Augmented Reality and Virtual Reality in Building Information Modeling: The Next Frontier in Civil Engineering Education. In G. Kurubacak & H. Altinpulluk (Eds.), *Mobile Technologies and Augmented Reality in Open Education* (pp. 233–261). Hershey, PA: IGI Global. doi:10.4018/978-1-5225-2110-5.ch012

Boulkroune, A. (2015). Fuzzy Adaptive Controller for Uncertain Multivariable Nonlinear Systems with Both Sector Nonlinearities and Dead-Zones. In A. Azar & S. Vaidyanathan (Eds.), *Handbook of Research on Advanced Intelligent Control Engineering and Automation* (pp. 334–363). Hershey, PA: IGI Global. doi:10.4018/978-1-4666-7248-2.ch012

Bozdogan, D., Kasap, B., & Kose, U. (2017). Design Principles for an Intelligent-Augmented-Reality-Based M-Learning Application to Improve Engineering Students' English Language Skills. In G. Kurubacak & H. Altinpulluk (Eds.), *Mobile Technologies and Augmented Reality in Open Education* (pp. 215–232). Hershey, PA: IGI Global. doi:10.4018/978-1-5225-2110-5.ch011

Breddermann, J., Martínez-Cerdá, J., & Torrent-Sellens, J. (2017). A Model for Teacher Training to Improve Students' 21st Century Skills in Online and Blended Learning: An Approach from Film Education. In N. Ostashewski, J. Howell, & M. Cleveland-Innes (Eds.), *Optimizing K-12 Education through Online and Blended Learning* (pp. 45–73). Hershey, PA: IGI Global. doi:10.4018/978-1-5225-0507-5.ch003

Carvalho, L. M. (2015). Challenges and Opportunities for Virtual Universities in the 21st Century. In P. Ordóñez de Pablos, R. Tennyson, & M. Lytras (Eds.), *Assessing the Role of Mobile Technologies and Distance Learning in Higher Education* (pp. 131–153). Hershey, PA: IGI Global. doi:10.4018/978-1-4666-7316-8.ch006

Cassard, A., & Sloboda, B. W. (2016). Faculty Perception of Virtual 3-D Learning Environment to Assess Student Learning. In D. Choi, A. Dailey-Hebert, & J. Simmons Estes (Eds.), *Emerging Tools and Applications of Virtual Reality in Education* (pp. 48–74). Hershey, PA: IGI Global. doi:10.4018/978-1-4666-9837-6.ch003

Charfeddine, M., Jouili, K., & Braiek, N. B. (2015). Approximate Input-Output Feedback Linearization of Non-Minimum Phase System using Vanishing Perturbation Theory. In A. Azar & S. Vaidyanathan (Eds.), *Handbook of Research on Advanced Intelligent Control Engineering and Automation* (pp. 173–201). Hershey, PA: IGI Global. doi:10.4018/978-1-4666-7248-2.ch006

Cinto, T., Leite, H. M., Carvalho, S. N., Peixoto, C. S., & Arantes, D. S. (2016). 3D Virtual Learning Environments: An Avatar-Based Virtual Classes Platform. In F. Neto, R. de Souza, & A. Gomes (Eds.), *Handbook of Research on 3-D Virtual Environments and Hypermedia for Ubiquitous Learning* (pp. 54–86). Hershey, PA: IGI Global. doi:10.4018/978-1-5225-0125-1.ch003

Clinton, V., Cooper, J. L., Michaelis, J. E., Alibali, M. W., & Nathan, M. J. (2017). How Revisions to Mathematical Visuals Affect Cognition: Evidence from Eye Tracking. In C. Was, F. Sansosti, & B. Morris (Eds.), *Eye-Tracking Technology Applications in Educational Research* (pp. 195–218). Hershey, PA: IGI Global. doi:10.4018/978-1-5225-1005-5.ch010

Cook, A. E., & Wei, W. (2017). Using Eye Movements to Study Reading Processes: Methodological Considerations. In C. Was, F. Sansosti, & B. Morris (Eds.), *Eye-Tracking Technology Applications in Educational Research* (pp. 27–47). Hershey, PA: IGI Global. doi:10.4018/978-1-5225-1005-5.ch002

Dames, L. S., Royal, C., & Sawyer-Kurian, K. M. (2017). Active Student Engagement through the Use of WebEx, MindTap, and a Residency Component to Teach a Masters Online Group Counseling Course. In J. Keengwe & P. Bull (Eds.), *Handbook of Research on Transformative Digital Content and Learning Technologies* (pp. 245–268). Hershey, PA: IGI Global. doi:10.4018/978-1-5225-2000-9.ch014

Das, S. K., Pota, H. R., & Petersen, I. R. (2015). Advanced Vibration Control of Atomic Force Microscope Scanner. In A. Azar & S. Vaidyanathan (Eds.), *Handbook of Research on Advanced Intelligent Control Engineering and Automation* (pp. 84–106). Hershey, PA: IGI Global. doi:10.4018/978-1-4666-7248-2.ch003

De Pasquale, D., Wood, E., Gottardo, A., Jones, J. A., Kaplan, R., & DeMarco, A. (2017). Tracking Children's Interactions with Traditional Text and Computer-Based Early Literacy Media. In C. Was, F. Sansosti, & B. Morris (Eds.), *Eye-Tracking Technology Applications in Educational Research* (pp. 107–121). Hershey, PA: IGI Global. doi:10.4018/978-1-5225-1005-5.ch006

Denkowski, M., & Sadkowski, Ł. (2015). Gesture-Driven System for Intelligent Building Control. In A. Azar & S. Vaidyanathan (Eds.), *Handbook of Research on Advanced Intelligent Control Engineering and Automation* (pp. 390–405). Hershey, PA: IGI Global. doi:10.4018/978-1-4666-7248-2.ch014

Desjarlais, M. (2017). The Use of Eye-gaze to Understand Multimedia Learning. In C. Was, F. Sansosti, & B. Morris (Eds.), *Eye-Tracking Technology Applications in Educational Research* (pp. 122–142). Hershey, PA: IGI Global. doi:10.4018/978-1-5225-1005-5.ch007

Dongre, S. S., & Malik, L. G. (2017). Data Stream Mining Using Ensemble Classifier: A Collaborative Approach of Classifiers. In V. Bhatnagar (Ed.), *Collaborative Filtering Using Data Mining and Analysis* (pp. 236–249). Hershey, PA: IGI Global. doi:10.4018/978-1-5225-0489-4.ch013

Dragnic-Cindric, D., Barrow, E., & Anderson, J. L. (2017). Opportunity to Start Strong: Integration of Technology in Science Lessons in the Early Elementary Grades. In J. Keengwe & P. Bull (Eds.), *Handbook of Research on Transformative Digital Content and Learning Technologies* (pp. 154–170). Hershey, PA: IGI Global. doi:10.4018/978-1-5225-2000-9.ch009

Eddine, O. H., Jamel, D., Selma, B. A., & Salah, S. (2015). Robust Iterative Learning Control for Linear Discrete-Time Switched Systems. In A. Azar & S. Vaidyanathan (Eds.), *Handbook of Research on Advanced Intelligent Control Engineering and Automation* (pp. 543–565). Hershey, PA: IGI Global. doi:10.4018/978-1-4666-7248-2.ch020

Ekren, G., & Keskın, N. O. (2017). Existing Standards and Programs for Use in Mobile Augmented Reality. In G. Kurubacak & H. Altinpulluk (Eds.), *Mobile Technologies and Augmented Reality in Open Education* (pp. 118–134). Hershey, PA: IGI Global. doi:10.4018/978-1-5225-2110-5.ch006

Emiroğlu, B. G., & Kurt, A. A. (2017). Use of Augmented Reality in Mobile Devices for Educational Purposes. In G. Kurubacak & H. Altinpulluk (Eds.), *Mobile Technologies and Augmented Reality in Open Education* (pp. 95–117). Hershey, PA: IGI Global. doi:10.4018/978-1-5225-2110-5.ch005

Estes, J. S., Dailey-Hebert, A., & Choi, D. H. (2016). Integrating Technological Innovations to Enhance the Teaching-Learning Process. In D. Choi, A. Dailey-Hebert, & J. Simmons Estes (Eds.), *Emerging Tools and Applications of Virtual Reality in Education* (pp. 277–304). Hershey, PA: IGI Global. doi:10.4018/978-1-4666-9837-6.ch013

Eutamene, A., Kholladi, M. K., Gaceb, D., & Belhadef, H. (2017). A Dual PSO-Adaptive Mean Shift for Preprocessing Optimization on Degraded Document Images. *International Journal of Applied Metaheuristic Computing*, *8*(1), 61–76. doi:10.4018/IJAMC.2017010104

Farmer, L. S. (2017). ICT Literacy Integration: Issues and Sample Efforts. In J. Keengwe & P. Bull (Eds.), *Handbook of Research on Transformative Digital Content and Learning Technologies* (pp. 59–80). Hershey, PA: IGI Global. doi:10.4018/978-1-5225-2000-9.ch004

Fernández, F. J., Jiménez, A. D., Manzano, F. S., & Márquez, J. M. (2017). An Energy Management Strategy and Fuel Cell Configuration Proposal for a Hybrid Renewable System with Hydrogen Backup. *International Journal of Energy Optimization and Engineering*, *6*(1), 1–22. doi:10.4018/IJEOE.2017010101

Ficarra, L. R., & Chapin, D. A. (2017). Reconceptualizing Universal Design for Learning (UDL) as Learning Technology in Non-Formal Education. In J. Keengwe & P. Bull (Eds.), *Handbook of Research on Transformative Digital Content and Learning Technologies* (pp. 81–102). Hershey, PA: IGI Global. doi:10.4018/978-1-5225-2000-9.ch005

Flor, B. P., & Flor, L. C. (2017). Authentic Assessment Construction in Online Education: The Case of the Open High School Program of the Philippines. In N. Ostashewski, J. Howell, & M. Cleveland-Innes (Eds.), *Optimizing K-12 Education through Online and Blended Learning* (pp. 225–239). Hershey, PA: IGI Global. doi:10.4018/978-1-5225-0507-5.ch012

Folk, J. R., & Eskenazi, M. A. (2017). Eye Movement Behavior and Individual Differences in Word Identification During Reading. In C. Was, F. Sansosti, & B. Morris (Eds.), *Eye-Tracking Technology Applications in Educational Research* (pp. 66–87). Hershey, PA: IGI Global. doi:10.4018/978-1-5225-1005-5.ch004

Freeman, E. L., Reyes, A. J., Dragnic-Cindric, D., & Anderson, J. L. (2017). Integrating Disciplinary Literacy Practices in One-to-One Classrooms. In J. Keengwe & P. Bull (Eds.), *Handbook of Research on Transformative Digital Content and Learning Technologies* (pp. 285–310). Hershey, PA: IGI Global. doi:10.4018/978-1-5225-2000-9.ch016

Ghanbarzadeh, R., & Ghapanchi, A. H. (2016). Applied Areas of Three Dimensional Virtual Worlds in Learning and Teaching: A Review of Higher Education. In D. Choi, A. Dailey-Hebert, & J. Simmons Estes (Eds.), *Emerging Tools and Applications of Virtual Reality in Education* (pp. 26–47). Hershey, PA: IGI Global. doi:10.4018/978-1-4666-9837-6.ch002

Gharbi, A., Gharsellaoui, H., & Khalgui, M. (2016). Real-Time Reconfigurations of Embedded Control Systems. *International Journal of System Dynamics Applications*, *5*(3), 71–93. doi:10.4018/IJSDA.2016070104

Ginoya, D., Shendge, P. D., & Phadke, S. B. (2014). Sliding mode control for mismatched uncertain systems using an extended disturbance observer. *Industrial Electronics, IEEE Transactions on*, *61*(4), 1983–1992. 10.1109/tie.2013.2271597

Goel, A., Rivera, W. A., Kincaid, P., Montgomery, M., Karwowski, W., & Finkelstein, N. M. (2016). Ethics in Virtual World Environments Research. In D. Choi, A. Dailey-Hebert, & J. Simmons Estes (Eds.), *Emerging Tools and Applications of Virtual Reality in Education* (pp. 258–276). Hershey, PA: IGI Global. doi:10.4018/978-1-4666-9837-6.ch012

Gouia, R., Gunn, C., & Audi, D. (2015). Using iPads in University Mathematics Classes: What Do the Students Think? In P. Ordóñez de Pablos, R. Tennyson, & M. Lytras (Eds.), *Assessing the Role of Mobile Technologies and Distance Learning in Higher Education* (pp. 60–77). Hershey, PA: IGI Global. doi:10.4018/978-1-4666-7316-8.ch003

Goyal, M., & Bhatnagar, V. (2017). A Classification Framework on Opinion Mining for Effective Recommendation Systems. In V. Bhatnagar (Ed.), *Collaborative Filtering Using Data Mining and Analysis* (pp. 180–194). Hershey, PA: IGI Global. doi:10.4018/978-1-5225-0489-4.ch010

Gritli, H., Belghith, S., & Khraief, N. (2015). Chaos Control of an Impact Mechanical Oscillator Based on the OGY Method. In A. Azar & S. Vaidyanathan (Eds.), *Handbook of Research on Advanced Intelligent Control Engineering and Automation* (pp. 259–278). Hershey, PA: IGI Global. doi:10.4018/978-1-4666-7248-2.ch009

Gritli, H., Khraief, N., & Belghith, S. (2015). Further Investigation of the Period-Three Route to Chaos in the Passive Compass-Gait Biped Model. In A. Azar & S. Vaidyanathan (Eds.), *Handbook of Research on Advanced Intelligent Control Engineering and Automation* (pp. 279–300). Hershey, PA: IGI Global. doi:10.4018/978-1-4666-7248-2.ch010

Guazzaroni, G. (2017). The Impact of Augmented Reality and Virtual Reality Study Material in the Future of Learning: A Teamwork Experience. In G. Kurubacak & H. Altinpulluk (Eds.), *Mobile Technologies and Augmented Reality in Open Education* (pp. 77–94). Hershey, PA: IGI Global. doi:10.4018/978-1-5225-2110-5.ch004

Guha, D., Roy, P. K., & Banerjee, S. (2016). Application of Modified Biogeography Based Optimization in AGC of an Interconnected Multi-Unit Multi-Source AC-DC Linked Power System. *International Journal of Energy Optimization and Engineering*, *5*(3), 1–18. doi:10.4018/IJEOE.2016070101

Gundogan, M. B. (2017). In Search for a "Good Fit" Between Augmented Reality and Mobile Learning Ecosystem. In G. Kurubacak & H. Altinpulluk (Eds.), *Mobile Technologies and Augmented Reality in Open Education* (pp. 135–153). Hershey, PA: IGI Global. doi:10.4018/978-1-5225-2110-5.ch007

H'mida, B., Sahbi, M., & Dhaou, S. (2015). Discrete-Time Approximation of Multivariable Continuous-Time Delay Systems. In A. Azar & S. Vaidyanathan (Eds.), *Handbook of Research on Advanced Intelligent Control Engineering and Automation* (pp. 516–542). Hershey, PA: IGI Global. doi:10.4018/978-1-4666-7248-2.ch019

Haghshenas, M., Sadeghzadeh, A., Shahbazi, R., & Nassiriyar, M. (2015). Mobile Wireless Technologies Application in Education. In P. Ordóñez de Pablos, R. Tennyson, & M. Lytras (Eds.), *Assessing the Role of Mobile Technologies and Distance Learning in Higher Education* (pp. 311–332). Hershey, PA: IGI Global. doi:10.4018/978-1-4666-7316-8.ch014

Hakim, D., Nabil, B., Mustapha, Z., Nacereddine, B., Haddad, S., & Mamar, S. A. (2015). Modelling and Realisation of a Three-Level PWM Inverter Using a DSP Controller to Feed an Asynchronous Machine. In A. Azar & S. Vaidyanathan (Eds.), *Handbook of Research on Advanced Intelligent Control Engineering and Automation* (pp. 687–718). Hershey, PA: IGI Global. doi:10.4018/978-1-4666-7248-2.ch025

Han, S. J., Liau-Hing, C., & Beyerlein, M. (2017). Facilitating Multicultural Student Team Engagement in Higher Education: A Model for Digital Learning Environments. In J. Keengwe & P. Bull (Eds.), *Handbook of Research on Transformative Digital Content and Learning Technologies* (pp. 184–210). Hershey, PA: IGI Global. doi:10.4018/978-1-5225-2000-9.ch011

Han, X., Fridman, E., & Spurgeon, S. K. (2012). Sliding mode control in the presence of input delay: A singular perturbation approach. *Automatica*, *48*(8), 1904–1912. doi:10.1016/j.automatica.2012.06.016

Hassan, A., & Jung, T. (2016). Augmented Reality as an Emerging Application in Tourism Marketing Education. In D. Choi, A. Dailey-Hebert, & J. Simmons Estes (Eds.), *Emerging Tools and Applications of Virtual Reality in Education* (pp. 168–185). Hershey, PA: IGI Global. doi:10.4018/978-1-4666-9837-6.ch008

Hayet, T., Hatem, T., & Jilani, K. (2015). Navigation Control of a Mobile Robot under Time Constraint using Genetic Algorithms, CSP Techniques, and Fuzzy Logic. In A. Azar & S. Vaidyanathan (Eds.), *Handbook of Research on Advanced Intelligent Control Engineering and Automation* (pp. 457–478). Hershey, PA: IGI Global. doi:10.4018/978-1-4666-7248-2.ch017

Hein, D. W., Jodoin, J. L., Rauschnabel, P. A., & Ivens, B. S. (2017). Are Wearables Good or Bad for Society?: An Exploration of Societal Benefits, Risks, and Consequences of Augmented Reality Smart Glasses. In G. Kurubacak & H. Altinpulluk (Eds.), *Mobile Technologies and Augmented Reality in Open Education* (pp. 1–25). Hershey, PA: IGI Global. doi:10.4018/978-1-5225-2110-5.ch001

Hrizi, O., Boussaid, B., Zouinkhi, A., & Abdelkrim, M. N. (2015). Robust Unknown Input Observer-Based Fast Adaptive Fault Estimation: Application to Mobile Robot. In A. Azar & S. Vaidyanathan (Eds.), *Handbook of Research on Advanced Intelligent Control Engineering and Automation* (pp. 427–456). Hershey, PA: IGI Global. doi:10.4018/978-1-4666-7248-2.ch016

Hur, J. W. (2017). Mobile Technology Integration and English Language Learners: A Case Study. In J. Keengwe & P. Bull (Eds.), *Handbook of Research on Transformative Digital Content and Learning Technologies* (pp. 25–41). Hershey, PA: IGI Global. doi:10.4018/978-1-5225-2000-9.ch002

Hussein, M. I. (2015). Evaluation of Mobile Learning Project at the UAE University: College of Engineering Case Study. In P. Ordóñez de Pablos, R. Tennyson, & M. Lytras (Eds.), *Assessing the Role of Mobile Technologies and Distance Learning in Higher Education* (pp. 100–130). Hershey, PA: IGI Global. doi:10.4018/978-1-4666-7316-8.ch005

Ireri, B. N., Wario, R. D., Omwenga, E. I., Oboko, R., & Mukiri, M. I. (2017). Mobile Learning: Content Format and Packaging for Effective Teaching and Learning in a Learner-Centered Pedagogy. In J. Keengwe & P. Bull (Eds.), *Handbook of Research on Transformative Digital Content and Learning Technologies* (pp. 329–344). Hershey, PA: IGI Global. doi:10.4018/978-1-5225-2000-9.ch018

Jain, A., Bhatnagar, V., & Sharma, P. (2017). Collaborative and Clustering Based Strategy in Big Data. In V. Bhatnagar (Ed.), *Collaborative Filtering Using Data Mining and Analysis* (pp. 140–158). Hershey, PA: IGI Global. doi:10.4018/978-1-5225-0489-4.ch008

Jones, G., & Alba, A. D. (2016). Reviewing the Effectiveness and Learning Outcomes of a 3D Virtual Museum: A Pilot Study. In F. Neto, R. de Souza, & A. Gomes (Eds.), *Handbook of Research on 3-D Virtual Environments and Hypermedia for Ubiquitous Learning* (pp. 168–191). Hershey, PA: IGI Global. doi:10.4018/978-1-5225-0125-1.ch007

Joshi, S., & Talange, D. B. (2016). Fault Tolerant Control of an AUV using Periodic Output Feedback with Multi Model Approach. *International Journal of System Dynamics Applications*, *5*(2), 41–62. doi:10.4018/IJSDA.2016040103

Kahenya, N. P. (2017). The Use of Social Media to Facilitate Real-Time eLearning. In J. Keengwe & P. Bull (Eds.), *Handbook of Research on Transformative Digital Content and Learning Technologies* (pp. 171–183). Hershey, PA: IGI Global. doi:10.4018/978-1-5225-2000-9.ch010

Kaleci, D., & Tepe, T. (2017). Design, Development, and Marketing Process of Video Games. In G. Kurubacak & H. Altinpulluk (Eds.), *Mobile Technologies and Augmented Reality in Open Education* (pp. 306–320). Hershey, PA: IGI Global. doi:10.4018/978-1-5225-2110-5.ch016

Kalsi, K., Lian, J., Hui, S., & Zak, S. H. (2010). Sliding-mode observers for systems with unknown inputs: A high-gain approach. *Automatica*, *46*(2), 347–353. doi:10.1016/j.automatica.2009.10.040

Kaplan, E., Levari, T., & Snedeker, J. (2017). Eye Tracking and Spoken Language Comprehension. In C. Was, F. Sansosti, & B. Morris (Eds.), *Eye-Tracking Technology Applications in Educational Research* (pp. 88–105). Hershey, PA: IGI Global. doi:10.4018/978-1-5225-1005-5.ch005

Katz, A., & Kim, J. H. (2017). Teaching Strategies and Tactics in K-12 Blended Education: The Flipped Classroom Model. In N. Ostashewski, J. Howell, & M. Cleveland-Innes (Eds.), *Optimizing K-12 Education through Online and Blended Learning* (pp. 156–184). Hershey, PA: IGI Global. doi:10.4018/978-1-5225-0507-5.ch009

Ke, S., & Lee, W. (2017). Combining User Co-Ratings and Social Trust for Collaborative Recommendation: A Data Analytics Approach. In V. Bhatnagar (Ed.), *Collaborative Filtering Using Data Mining and Analysis* (pp. 195–216). Hershey, PA: IGI Global. doi:10.4018/978-1-5225-0489-4.ch011

Kermani, M., & Sakly, A. (2015). On Stability Analysis of Switched Linear Time-Delay Systems under Arbitrary Switching. In A. Azar & S. Vaidyanathan (Eds.), *Handbook of Research on Advanced Intelligent Control Engineering and Automation* (pp. 480–515). Hershey, PA: IGI Global. doi:10.4018/978-1-4666-7248-2.ch018

Khalique, M., & Tunggau, S. L. (2015). Factors Influencing Behavior of Selecting Touch Screen Mobile Phones. In P. Ordóñez de Pablos, R. Tennyson, & M. Lytras (Eds.), *Assessing the Role of Mobile Technologies and Distance Learning in Higher Education* (pp. 297–310). Hershey, PA: IGI Global. doi:10.4018/978-1-4666-7316-8.ch013

Kharola, A., & Patil, P. P. (2017). Neural Fuzzy Control of Ball and Beam System. *International Journal of Energy Optimization and Engineering*, *6*(2), 64–78. doi:10.4018/IJEOE.2017040104

Kim, J. H., Foster, A., & Cho, M. (2017). Professional Development for Technology Integration into Differentiated Math Instruction. In J. Keengwe & P. Bull (Eds.), *Handbook of Research on Transformative Digital Content and Learning Technologies* (pp. 1–24). Hershey, PA: IGI Global. doi:10.4018/978-1-5225-2000-9.ch001

Kithinji, W. K., & Kanga, A. W. (2017). Distance Learning in Kenyan Universities: The Relationship between Learners' Characteristics and Academic Performance. In J. Keengwe & P. Bull (Eds.), *Handbook of Research on Transformative Digital Content and Learning Technologies* (pp. 231–244). Hershey, PA: IGI Global. doi:10.4018/978-1-5225-2000-9.ch013

Koole, M., Dionne, J., McCoy, E. T., & Epp, J. (2017). Makerspaces: Materializing, Digitizing, and Transforming Learning. In J. Keengwe & P. Bull (Eds.), *Handbook of Research on Transformative Digital Content and Learning Technologies* (pp. 211–230). Hershey, PA: IGI Global. doi:10.4018/978-1-5225-2000-9.ch012

Korres, M. P. (2015). Promoting Interaction in an Asynchronous E-Learning Environment. In P. Ordóñez de Pablos, R. Tennyson, & M. Lytras (Eds.), *Assessing the Role of Mobile Technologies and Distance Learning in Higher Education* (pp. 154–175). Hershey, PA: IGI Global. doi:10.4018/978-1-4666-7316-8.ch007

Köse, U. (2017). An Augmented-Reality-Based Intelligent Mobile Application for Open Computer Education. In G. Kurubacak & H. Altinpulluk (Eds.), *Mobile Technologies and Augmented Reality in Open Education* (pp. 154–174). Hershey, PA: IGI Global. doi:10.4018/978-1-5225-2110-5.ch008

Leighton, L. J., & Crompton, H. (2017). Augmented Reality in K-12 Education. In G. Kurubacak & H. Altinpulluk (Eds.), *Mobile Technologies and Augmented Reality in Open Education* (pp. 281–290). Hershey, PA: IGI Global. doi:10.4018/978-1-5225-2110-5.ch014

Leung, C. K., Jiang, F., Dela Cruz, E. M., & Elango, V. S. (2017). Association Rule Mining in Collaborative Filtering. In V. Bhatnagar (Ed.), *Collaborative Filtering Using Data Mining and Analysis* (pp. 159–179). Hershey, PA: IGI Global. doi:10.4018/978-1-5225-0489-4.ch009

Levesque, A., & Reid, D. (2017). Factors Influencing International Student Success in a K-12 Blended Learning Program. In N. Ostashewski, J. Howell, & M. Cleveland-Innes (Eds.), *Optimizing K-12 Education through Online and Blended Learning* (pp. 93–108). Hershey, PA: IGI Global. doi:10.4018/978-1-5225-0507-5.ch005

Liu, M., Su, S., Liu, S., Harron, J., Fickert, C., & Sherman, B. (2016). Exploring 3D Immersive and Interactive Technology for Designing Educational Learning Experiences. In F. Neto, R. de Souza, & A. Gomes (Eds.), *Handbook of Research on 3-D Virtual Environments and Hypermedia for Ubiquitous Learning* (pp. 243–261). Hershey, PA: IGI Global. doi:10.4018/978-1-5225-0125-1.ch010

Liu, Z., Cheng, H. N., Liu, S., & Sun, J. (2017). Discovering the Two-Step Lag Behavioral Patterns of Learners in the College SPOC Platform. *International Journal of Information and Communication Technology Education, 13*(1), 1–13. doi:10.4018/IJICTE.2017010101

Ludlow, B. L., & Hartley, M. D. (2016). Using Second Life® for Situated and Active Learning in Teacher Education. In D. Choi, A. Dailey-Hebert, & J. Simmons Estes (Eds.), *Emerging Tools and Applications of Virtual Reality in Education* (pp. 96–120). Hershey, PA: IGI Global. doi:10.4018/978-1-4666-9837-6.ch005

Luo, L., Kiewra, K. A., Peteranetz, M. S., & Flanigan, A. E. (2017). Using Eye-Tracking Technology to Understand How Graphic Organizers Aid Student Learning. In C. Was, F. Sansosti, & B. Morris (Eds.), *Eye-Tracking Technology Applications in Educational Research* (pp. 220–238). Hershey, PA: IGI Global. doi:10.4018/978-1-5225-1005-5.ch011

M., V., & K., T. (2017). History and Overview of the Recommender Systems. In V. Bhatnagar (Ed.), *Collaborative Filtering Using Data Mining and Analysis* (pp. 74-99). Hershey, PA: IGI Global. doi:10.4018/978-1-5225-0489-4.ch004

Mahajan, R. (2017). Review of Data Mining Techniques and Parameters for Recommendation of Effective Adaptive E-Learning System. In V. Bhatnagar (Ed.), *Collaborative Filtering Using Data Mining and Analysis* (pp. 1–23). Hershey, PA: IGI Global. doi:10.4018/978-1-5225-0489-4.ch001

Mahmoud, M. S., & Qureshi, A. U. D. (2012). Decentralized sliding-mode output-feedback control of interconnected discrete-delay systems. *Automatica, 48*(5), 808–814. doi:10.1016/j.automatica.2012.02.008

Margitay-Becht, A. (2016). Teaching Economics in World of Warcraft. In D. Choi, A. Dailey-Hebert, & J. Simmons Estes (Eds.), *Emerging Tools and Applications of Virtual Reality in Education* (pp. 121–144). Hershey, PA: IGI Global. doi:10.4018/978-1-4666-9837-6.ch006

Marinakou, E., & Giousmpasoglou, C. (2015). M-Learning in the Middle East: The Case of Bahrain. In P. Ordóñez de Pablos, R. Tennyson, & M. Lytras (Eds.), *Assessing the Role of Mobile Technologies and Distance Learning in Higher Education* (pp. 176–199). Hershey, PA: IGI Global. doi:10.4018/978-1-4666-7316-8.ch008

McVey, M. K., Poyo, S., & Smith, M. L. (2017). Optimizing K-12 Education through Effective Educator Preparation: Lessons Learned from a Synchronous Online Pilot Study. In N. Ostashewski, J. Howell, & M. Cleveland-Innes (Eds.), *Optimizing K-12 Education through Online and Blended Learning* (pp. 23–44). Hershey, PA: IGI Global. doi:10.4018/978-1-5225-0507-5.ch002

Meshur, H. F., & Bala, H. A. (2015). Distance Learning in Architecture/Planning Education: A Case Study in the Faculty of Architecture at Selcuk University. In P. Ordóñez de Pablos, R. Tennyson, & M. Lytras (Eds.), *Assessing the Role of Mobile Technologies and Distance Learning in Higher Education* (pp. 1–28). Hershey, PA: IGI Global. doi:10.4018/978-1-4666-7316-8.ch001

Miladi, Y., & Feki, M. (2015). Bifurcation, Quasi-Periodicity, Chaos, and Co-Existence of Different Behaviors in the Controlled H-Bridge Inverter. In A. Azar & S. Vaidyanathan (Eds.), *Handbook of Research on Advanced Intelligent Control Engineering and Automation* (pp. 301–332). Hershey, PA: IGI Global. doi:10.4018/978-1-4666-7248-2.ch011

Minshew, L. M., & Anderson, J. L. (2017). Integrating iPads in Middle School Science Instruction: A Case Study. In J. Keengwe & P. Bull (Eds.), *Handbook of Research on Transformative Digital Content and Learning Technologies* (pp. 42–58). Hershey, PA: IGI Global. doi:10.4018/978-1-5225-2000-9.ch003

Mittal, M., Sharma, R. K., Singh, V., & Mohan Goyal, L. (2017). Modified Single Pass Clustering Algorithm Based on Median as a Threshold Similarity Value. In V. Bhatnagar (Ed.), *Collaborative Filtering Using Data Mining and Analysis* (pp. 24–48). Hershey, PA: IGI Global. doi:10.4018/978-1-5225-0489-4.ch002

Mkrttchian, V., Amirov, D., & Belyanina, L. (2017). Optimizing an Online Learning Course Using Automatic Curating in Sliding Mode. In N. Ostashewski, J. Howell, & M. Cleveland-Innes (Eds.), *Optimizing K-12 Education through Online and Blended Learning* (pp. 213–224). Hershey, PA: IGI Global. doi:10.4018/978-1-5225-0507-5.ch011

Moysis, L., & Azar, A. T. (2017). New Discrete Time 2D Chaotic Maps. *International Journal of System Dynamics Applications*, *6*(1), 77–104. doi:10.4018/IJSDA.2017010105

Nazareth, A., Odean, R., & Pruden, S. M. (2017). The Use of Eye-Tracking in Spatial Thinking Research. In C. Was, F. Sansosti, & B. Morris (Eds.), *Eye-Tracking Technology Applications in Educational Research* (pp. 239–260). Hershey, PA: IGI Global. doi:10.4018/978-1-5225-1005-5.ch012

Newton, R. R. (2017). Fully Online Education and Underserved Populations. In J. Keengwe & P. Bull (Eds.), *Handbook of Research on Transformative Digital Content and Learning Technologies* (pp. 118–136). Hershey, PA: IGI Global. doi:10.4018/978-1-5225-2000-9.ch007

Nishani, L., & Biba, M. (2017). Statistical Relational Learning for Collaborative Filtering a State-of-the-Art Review. In V. Bhatnagar (Ed.), *Collaborative Filtering Using Data Mining and Analysis* (pp. 250–269). Hershey, PA: IGI Global. doi:10.4018/978-1-5225-0489-4.ch014

O'Connor, A., Seery, N., & Canty, D. (2017). The Psychological Domain: Enhancing Traditional Practice in K-12 Education. In N. Ostashewski, J. Howell, & M. Cleveland-Innes (Eds.), *Optimizing K-12 Education through Online and Blended Learning* (pp. 109–127). Hershey, PA: IGI Global. doi:10.4018/978-1-5225-0507-5.ch006

Oliveira, T. A., Marranghello, N., Silva, A. C., & Pereira, A. S. (2016). Virtual Laboratories Development Using 3D Environments. In F. Neto, R. de Souza, & A. Gomes (Eds.), *Handbook of Research on 3-D Virtual Environments and Hypermedia for Ubiquitous Learning* (pp. 29–53). Hershey, PA: IGI Global. doi:10.4018/978-1-5225-0125-1.ch002

Ossiannilsson, E. (2017). Leadership in Global Open, Online, and Distance Learning. In J. Keengwe & P. Bull (Eds.), *Handbook of Research on Transformative Digital Content and Learning Technologies* (pp. 345–373). Hershey, PA: IGI Global. doi:10.4018/978-1-5225-2000-9.ch019

Özdemir, M. (2017). Educational Augmented Reality (AR) Applications and Development Process. In G. Kurubacak & H. Altinpulluk (Eds.), *Mobile Technologies and Augmented Reality in Open Education* (pp. 26–53). Hershey, PA: IGI Global. doi:10.4018/978-1-5225-2110-5.ch002

Pal, A., & Kumar, M. (2017). Collaborative Filtering Based Data Mining for Large Data. In V. Bhatnagar (Ed.), *Collaborative Filtering Using Data Mining and Analysis* (pp. 115–127). Hershey, PA: IGI Global. doi:10.4018/978-1-5225-0489-4.ch006

Paliktzoglou, V., Stylianou, T., & Suhonen, J. (2015). Google Educational Apps as a Collaborative Learning Tool among Computer Science Learners. In P. Ordóñez de Pablos, R. Tennyson, & M. Lytras (Eds.), *Assessing the Role of Mobile Technologies and Distance Learning in Higher Education* (pp. 272–296). Hershey, PA: IGI Global. doi:10.4018/978-1-4666-7316-8.ch012

Pandey, K. (2015). Mobile Education Mitigating the Heavy Magnitude of Illiteracy in India. In P. Ordóñez de Pablos, R. Tennyson, & M. Lytras (Eds.), *Assessing the Role of Mobile Technologies and Distance Learning in Higher Education* (pp. 200–227). Hershey, PA: IGI Global. doi:10.4018/978-1-4666-7316-8.ch009

Parikh, C. (2017). Eye-Tracking Technology: A Closer Look at Eye-Tracking Paradigms with High-Risk Populations. In C. Was, F. Sansosti, & B. Morris (Eds.), *Eye-Tracking Technology Applications in Educational Research* (pp. 283–302). Hershey, PA: IGI Global. doi:10.4018/978-1-5225-1005-5.ch014

Patterson, R. L., Patterson, D. C., & Robertson, A. (2016). Seeing Numbers Differently: Mathematics in the Virtual World. In D. Choi, A. Dailey-Hebert, & J. Simmons Estes (Eds.), *Emerging Tools and Applications of Virtual Reality in Education* (pp. 186–214). Hershey, PA: IGI Global. doi:10.4018/978-1-4666-9837-6.ch009

Pham, V., Volos, C., & Vaidyanathan, S. (2015). Chaotic Attractor in a Novel Time-Delayed System with a Saturation Function. In A. Azar & S. Vaidyanathan (Eds.), *Handbook of Research on Advanced Intelligent Control Engineering and Automation* (pp. 230–258). Hershey, PA: IGI Global. doi:10.4018/978-1-4666-7248-2.ch008

Pigatt, Y., & Braman, J. (2016). Increasing Student Engagement through Virtual Worlds: A Community College Approach in a Diversity Course. In D. Choi, A. Dailey-Hebert, & J. Simmons Estes (Eds.), *Emerging Tools and Applications of Virtual Reality in Education* (pp. 75–94). Hershey, PA: IGI Global. doi:10.4018/978-1-4666-9837-6.ch004

Polly, D. (2015). Leveraging Asynchronous Online Instruction to Develop Elementary School Mathematics Teacher-Leaders. In P. Ordóñez de Pablos, R. Tennyson, & M. Lytras (Eds.), *Assessing the Role of Mobile Technologies and Distance Learning in Higher Education* (pp. 78–99). Hershey, PA: IGI Global. doi:10.4018/978-1-4666-7316-8.ch004

Rochecouste, J., & Oliver, R. (2017). Introducing the Teaching and Learning Benefits of the WWW in Aboriginal Schools: Trials and Tribulations. In N. Ostashewski, J. Howell, & M. Cleveland-Innes (Eds.), *Optimizing K-12 Education through Online and Blended Learning* (pp. 128–137). Hershey, PA: IGI Global. doi:10.4018/978-1-5225-0507-5.ch007

Rodrigues, P., & Rosa, P. J. (2017). Eye-Tracking as a Research Methodology in Educational Context: A Spanning Framework. In C. Was, F. Sansosti, & B. Morris (Eds.), *Eye-Tracking Technology Applications in Educational Research* (pp. 1–26). Hershey, PA: IGI Global. doi:10.4018/978-1-5225-1005-5.ch001

Royal, C., Wasik, S., Horne, R., Dames, L. S., & Newsome, G. (2017). Digital Wellness: Integrating Wellness in Everyday Life with Digital Content and Learning Technologies. In J. Keengwe & P. Bull (Eds.), *Handbook of Research on Transformative Digital Content and Learning Technologies* (pp. 103–117). Hershey, PA: IGI Global. doi:10.4018/978-1-5225-2000-9.ch006

Saini, A. (2017). Big Data Mining Using Collaborative Filtering. In V. Bhatnagar (Ed.), *Collaborative Filtering Using Data Mining and Analysis* (pp. 128–138). Hershey, PA: IGI Global. doi:10.4018/978-1-5225-0489-4.ch007

Sala, N. M. (2016). Virtual Reality and Education: Overview Across Different Disciplines. In D. Choi, A. Dailey-Hebert, & J. Simmons Estes (Eds.), *Emerging Tools and Applications of Virtual Reality in Education* (pp. 1–25). Hershey, PA: IGI Global. doi:10.4018/978-1-4666-9837-6.ch001

Saleeb, N., Dafoulas, G. A., & Loomes, M. (2016). Personalisation of 3D Virtual Spaces for Enhanced Ubiquitous Learning. In F. Neto, R. de Souza, & A. Gomes (Eds.), *Handbook of Research on 3-D Virtual Environments and Hypermedia for Ubiquitous Learning* (pp. 87–114). Hershey, PA: IGI Global. doi:10.4018/978-1-5225-0125-1.ch004

Salhi, H., & Kamoun, S. (2015). State and Parametric Estimation of Nonlinear Systems Described by Wiener Sate-Space Mathematical Models. In A. Azar & S. Vaidyanathan (Eds.), *Handbook of Research on Advanced Intelligent Control Engineering and Automation* (pp. 107–145). Hershey, PA: IGI Global. doi:10.4018/978-1-4666-7248-2.ch004

Salifu, S. (2017). A Blueprint for Online Licensed Practical Nurse Training. In J. Keengwe & P. Bull (Eds.), *Handbook of Research on Transformative Digital Content and Learning Technologies* (pp. 374–392). Hershey, PA: IGI Global. doi:10.4018/978-1-5225-2000-9.ch020

Salinas, P. (2017). Augmented Reality: Opportunity for Developing Spatial Visualization and Learning Calculus. In G. Kurubacak & H. Altinpulluk (Eds.), *Mobile Technologies and Augmented Reality in Open Education* (pp. 54–76). Hershey, PA: IGI Global. doi:10.4018/978-1-5225-2110-5.ch003

Sangwan, N., & Dahiya, N. (2017). A Classification Framework Towards Application of Data Mining in Collaborative Filtering. In V. Bhatnagar (Ed.), *Collaborative Filtering Using Data Mining and Analysis* (pp. 100–114). Hershey, PA: IGI Global. doi:10.4018/978-1-5225-0489-4.ch005

Scheiter, K., & Eitel, A. (2017). The Use of Eye Tracking as a Research and Instructional Tool in Multimedia Learning. In C. Was, F. Sansosti, & B. Morris (Eds.), *Eye-Tracking Technology Applications in Educational Research* (pp. 143–164). Hershey, PA: IGI Global. doi:10.4018/978-1-5225-1005-5.ch008

Schroeder, N. L. (2016). Pedagogical Agents for Learning. In D. Choi, A. Dailey-Hebert, & J. Simmons Estes (Eds.), *Emerging Tools and Applications of Virtual Reality in Education* (pp. 216–238). Hershey, PA: IGI Global. doi:10.4018/978-1-4666-9837-6.ch010

Scott, D. M. (2017). Using Hearing Assistance Technology to Improve School Success for All Children. In J. Keengwe & P. Bull (Eds.), *Handbook of Research on Transformative Digital Content and Learning Technologies* (pp. 311–328). Hershey, PA: IGI Global. doi:10.4018/978-1-5225-2000-9.ch017

Shayan, S., Abrahamson, D., Bakker, A., Duijzer, C. A., & van der Schaaf, M. (2017). Eye-Tracking the Emergence of Attentional Anchors in a Mathematics Learning Tablet Activity. In C. Was, F. Sansosti, & B. Morris (Eds.), *Eye-Tracking Technology Applications in Educational Research* (pp. 166–194). Hershey, PA: IGI Global. doi:10.4018/978-1-5225-1005-5.ch009

Sheffield, R., & Quinton, G. (2017). Case Studies of Scaffolded On-Line Inquiry in Primary and Secondary Classrooms: Technology and Inquiry in a Science Context. In N. Ostashewski, J. Howell, & M. Cleveland-Innes (Eds.), *Optimizing K-12 Education through Online and Blended Learning* (pp. 240–255). Hershey, PA: IGI Global. doi:10.4018/978-1-5225-0507-5.ch013

Stenbom, S., Cleveland-Innes, M., & Hrastinski, S. (2017). Online Coaching as Teacher Training: Using a Relationship of Inquiry Framework. In N. Ostashewski, J. Howell, & M. Cleveland-Innes (Eds.), *Optimizing K-12 Education through Online and Blended Learning* (pp. 1–22). Hershey, PA: IGI Global. doi:10.4018/978-1-5225-0507-5.ch001

Sternig, C., Spitzer, M., & Ebner, M. (2017). Learning in a Virtual Environment: Implementation and Evaluation of a VR Math-Game. In G. Kurubacak & H. Altinpulluk (Eds.), *Mobile Technologies and Augmented Reality in Open Education* (pp. 175–199). Hershey, PA: IGI Global. doi:10.4018/978-1-5225-2110-5.ch009

Sural, I. (2017). Mobile Augmented Reality Applications in Education. In G. Kurubacak & H. Altinpulluk (Eds.), *Mobile Technologies and Augmented Reality in Open Education* (pp. 200–214). Hershey, PA: IGI Global. doi:10.4018/978-1-5225-2110-5.ch010

Tan, C. P., Yu, X., & Man, Z. (2010). Terminal sliding mode observers for a class of nonlinear systems. *Automatica*, *46*(8), 1401–1404. doi:10.1016/j.automatica.2010.05.010

Tavernise, A., & Bertacchini, F. (2016). Designing Educational Paths in Virtual Worlds for a Successful Hands-On Learning: Cultural Scenarios in NetConnect Project. In F. Neto, R. de Souza, & A. Gomes (Eds.), *Handbook of Research on 3-D Virtual Environments and Hypermedia for Ubiquitous Learning* (pp. 148–167). Hershey, PA: IGI Global. doi:10.4018/978-1-5225-0125-1.ch006

Taylan, R. D. (2017). Promoting Active Learning in Mathematics Teacher Education: The Flipped Classroom Method and Use of Video Content. In J. Keengwe & P. Bull (Eds.), *Handbook of Research on Transformative Digital Content and Learning Technologies* (pp. 269–284). Hershey, PA: IGI Global. doi:10.4018/978-1-5225-2000-9.ch015

Thuku, J. K., Maina, E. M., Ondigi, S. R., & Ayot, H. O. (2017). Enhancing Learner-Centered Instruction through Tutorial Management Using Cloud Computing. In J. Keengwe & P. Bull (Eds.), *Handbook of Research on Transformative Digital Content and Learning Technologies* (pp. 137–153). Hershey, PA: IGI Global. doi:10.4018/978-1-5225-2000-9.ch008

Todinov, M. (2017). Reducing Risk Through Inversion and Self-Strengthening. *International Journal of Risk and Contingency Management*, *6*(1), 14–42. doi:10.4018/IJRCM.2017010102

Tosun, N. (2017). Augmented Reality Implementations, Requirements, and Limitations in the Flipped-Learning Approach. In G. Kurubacak & H. Altinpulluk (Eds.), *Mobile Technologies and Augmented Reality in Open Education* (pp. 262–280). Hershey, PA: IGI Global. doi:10.4018/978-1-5225-2110-5.ch013

Tsai, W., & Ma, C. (2017). Automatic Identification of Simultaneous and Non-Simultaneous Singers for Music Data Indexing. *International Journal of Web Services Research*, *14*(1), 29–43. doi:10.4018/IJWSR.2017010103

Uluyol, Ç., & Şahin, S. (2016). Augmented Reality: A New Direction in Education. In D. Choi, A. Dailey-Hebert, & J. Simmons Estes (Eds.), *Emerging Tools and Applications of Virtual Reality in Education* (pp. 239–257). Hershey, PA: IGI Global. doi:10.4018/978-1-4666-9837-6.ch011

Vaughan, N. (2017). An Inquiry-Based Approach to Blended and Online Learning in K-12 Education. In N. Ostashewski, J. Howell, & M. Cleveland-Innes (Eds.), *Optimizing K-12 Education through Online and Blended Learning* (pp. 138–155). Hershey, PA: IGI Global. doi:10.4018/978-1-5225-0507-5.ch008

Volos, C., Kyprianidis, I., Stouboulos, I., & Vaidyanathan, S. (2015). Random Bit Generator Based on Non-Autonomous Chaotic Systems. In A. Azar & S. Vaidyanathan (Eds.), *Handbook of Research on Advanced Intelligent Control Engineering and Automation* (pp. 203–229). Hershey, PA: IGI Global. doi:10.4018/978-1-4666-7248-2.ch007

Volos, C., Kyprianidis, I., Stouboulos, I., & Vaidyanathan, S. (2016). Design of a Chaotic Random Bit Generator Using a Duffing - van der Pol System. *International Journal of System Dynamics Applications*, *5*(3), 94–111. doi:10.4018/IJSDA.2016070105

Yasuda, G. (2015). Distributed Coordination Architecture for Cooperative Task Planning and Execution of Intelligent Multi-Robot Systems. In A. Azar & S. Vaidyanathan (Eds.), *Handbook of Research on Advanced Intelligent Control Engineering and Automation* (pp. 407–426). Hershey, PA: IGI Global. doi:10.4018/978-1-4666-7248-2.ch015

Yeh, E., & Wan, G. (2016). The Use of Virtual Worlds in Foreign Language Teaching and Learning. In D. Choi, A. Dailey-Hebert, & J. Simmons Estes (Eds.), *Emerging Tools and Applications of Virtual Reality in Education* (pp. 145–167). Hershey, PA: IGI Global. doi:10.4018/978-1-4666-9837-6.ch007

Yuksekdag, B. B. (2017). The Importance of Mobile Augmented Reality in Online Nursing Education. In G. Kurubacak & H. Altinpulluk (Eds.), *Mobile Technologies and Augmented Reality in Open Education* (pp. 291–305). Hershey, PA: IGI Global. doi:10.4018/978-1-5225-2110-5.ch015

Zaghdoud, R., Salhi, S., & Ksouri, M. (2015). On Proportional Plus Derivative State Feedback H2 Control for Descriptor Systems. In A. Azar & S. Vaidyanathan (Eds.), *Handbook of Research on Advanced Intelligent Control Engineering and Automation* (pp. 146–172). Hershey, PA: IGI Global. doi:10.4018/978-1-4666-7248-2.ch005

Zamora, R., Vélez, J., & Villa, J. L. (2016). Contributions of Collaborative and Immersive Environments in Development a Remote Access Laboratory: From Point of View of Effectiveness in Learning. In F. Neto, R. de Souza, & A. Gomes (Eds.), *Handbook of Research on 3-D Virtual Environments and Hypermedia for Ubiquitous Learning* (pp. 1–28). Hershey, PA: IGI Global. doi:10.4018/978-1-5225-0125-1.ch001

Zentall, S. R., & Junglen, A. G. (2017). Investigating Mindsets and Motivation through Eye Tracking and Other Physiological Measures. In C. Was, F. Sansosti, & B. Morris (Eds.), *Eye-Tracking Technology Applications in Educational Research* (pp. 48–64). Hershey, PA: IGI Global. doi:10.4018/978-1-5225-1005-5.ch003

About the Authors

Vardan Mkrttchian received his Doctorate of Sciences (Engineering) in Control Systems from Lomonosov Moscow State University (former USSR). Dr. Vardan Mkrttchian taught undergraduate and graduate information and control system courses at the Astrakhan State University (Russian Federation) science 2010, where he is the Professor of the Information and Control Systems Department (www.aspu.ru). He is currently chief executive and rector and leader of the international academic scientist's team of HHH University, Australia, (www.hhhuniversity.com). He also serves as executive director of the HHH Technology Incorporation. Professor Vardan Mkrttchian has authored over 350 refereed publications He is the author of over twenty books in information technology, control system theory, electronics, and cloud and virtual education technology. He also has authored more than 190 articles published in various conference proceedings and journals.

Ekaterina Aleshina graduated from Penza State Pedagogical University, the Faculty of Foreign Languages with English as a major and German as a minor (Philology), getting qualifications of a teacher of English and German. E. Aleshina got her PhD in History from the above university in 2006 followed by the title of associate professor in 2008. In 2009-2010, E. Aleshina was doing Fulbright Faculty Development Program in Southern Connecticut State University (USA). While in the USA, she was invited as guest speaker to several universities and high schools. In 2010-2014, E. Aleshina was head of department of English and English Language Teaching Methodology at Penza State Pedagogical University, now she is head of department of Foreign Languages and FLT Methodology at Penza State University. Currently, E. Aleshina is doing her doctoral research in the specificity of political communication at Moscow Pedagogical State University. She has authored about 50 publication featuring interdisciplinary issues and approaches.

Index

T

V

Printed in the United States
By Bookmasters